PRAISE FOR *Carbon Shock*

"With his skill as a writer and long experience as an investigative reporter, Mark Schapiro brings alive some unexpected angles of the most important story of our time. I thought I knew the basics of carbon and climate change, but reading this lively and intriguing book made me aware of much I didn't know—both fascinating and disturbing."

—Adam Hochschild, author of *To End All Wars*

"Can keepers of the fossil flames ever be persuaded that we're all imperiled if we don't de-carbonize? Is it delusional to imagine building monetary bridges to a cleaner future, so that civilization—at least the civil parts of it, including everyone's job—might survive if we did? Is any government actually still in charge, as we face what's surely humanity's greatest challenge? We can be grateful that Mark Schapiro has navigated some dreaded territory—the arcana of global finance—to show with blessed clarity exactly where we are so far, what's failed and why, what might work, and where surprising hope lies."

—Alan Weisman, author of *Gaviotas, The World Without Us*, and
Countdown: Our Last, Best Hope for a Future on Earth?

"In his powerful new book, *Carbon Shock*, Mark Schapiro transcends standard discussions about the well-known culprits and ramifications of climate change and takes us on a harrowing, international exploration of the universal economic costs of carbon emissions. In his path-breaking treatise, Schapiro exposes the multinational corporate obfuscation of these costs; the folly of localized pseudosolutions that spur Wall Street trading but don't quantify financial costs or public risks, solve core problems, or provide socially cheaper and environmentally sounder practices; and the laggard policies of the US, Russia, and China relative to the EU in fashioning longer-term remedies. Not only does Schapiro compel the case for a global effort to thwart the joint economic and environmental plundering of our planet in this formidable book, but he expertly outlines the way to get there."

—Nomi Prins, author of *All the Presidents'*
Bankers and *It Takes a Pillage*

CARBON SHOCK

CARBON SHOCK
A Tale of Risk and Calculus on the Front Lines of the Disrupted Global Economy

How Carbon Is Changing
the Cost of Everything

MARK SCHAPIRO

CHELSEA GREEN PUBLISHING
WHITE RIVER JUNCTION, VERMONT

Editor: Joni Praded
Project Manager: Bill Bokermann
Copy Editor: Laura Jorstad
Proofreader: Helen Walden
Indexer: Barbara Mortenson

Printed in the United States of America.
First printing August, 2014
10 9 8 7 6 5 4 3 2 1   14 15 16 17

green press INITIATIVE

Chelsea Green Publishing is committed to preserving ancient forests and natural resources. We elected
to print this title on paper containing 100% postconsumer recycled paper, processed chlorine-free. As a
result, for this printing, we have saved:

71 Trees (40' tall and 6-8" diameter)
33,202 Gallons of Wastewater
32 million BTUs Total Energy
2,223 Pounds of Solid Waste
6,122 Pounds of Greenhouse Gases

Chelsea Green Publishing made this paper choice
because we are a member of the Green Press Initiative,
a nonprofit program dedicated to supporting authors,
publishers, and suppliers in their efforts to reduce their
use of fiber obtained from endangered forests. For
more information, visit www.greenpressinitiative.org.

Environmental impact estimates were made using
the Environmental Defense Paper Calculator. For
more information visit: www.papercalculator.org.

Our Commitment to Green Publishing
Chelsea Green sees publishing as a tool for cultural change and ecological stewardship. We strive
to align our book manufacturing practices with our editorial mission and to reduce the impact
of our business enterprise in the environment. We print our books and catalogs on chlorine-
free recycled paper, using vegetable-based inks whenever possible. This book may cost slightly
more because it was printed on paper that contains recycled fiber, and we hope you'll agree that
it's worth it. Chelsea Green is a member of the Green Press Initiative (www.greenpressinitiative
.org), a nonprofit coalition of publishers, manufacturers, and authors working to protect the
world's endangered forests and conserve natural resources. *Carbon Shock* was printed on paper
supplied by Maple Press that contains 100% postconsumer recycled fiber.

Library of Congress Cataloging-in-Publication Data
Schapiro, Mark, 1955–
  Carbon shock : a tale of risk and calculus on the front lines of the disrupted global
economy / Mark Schapiro.
     pages cm
  Includes bibliographical references and index.
  ISBN 978-1-60358-557-6 (hardback)—ISBN 978-1-60358-558-3 (ebook) 1. Environmental
economics. 2. Energy industries—Environmental aspects. 3. Economic policy—
Environmental aspects. 4. Carbon dioxide mitigation. I. Title.
  HC79.E5S28257 2014
  363.738'74—dc23
                                   2014017467

Chelsea Green Publishing
85 North Main Street, Suite 120
White River Junction, VT 05001
(802) 295-6300
www.chelseagreen.com

MIX
Paper from
responsible sources
FSC® C068106

To my brothers, Erik and Seth.

CONTENTS

INTRODUCTION

C haos and uncertainty. Those are the characteristics of the natural world under the pressures being wrought by climate change. Normal patterns of rainfall, temperature, and extreme weather are changing so rapidly that past baselines for these primordial forces are less and less relevant.

Scientists have a word for this state of flux. They call it "the end of stationarity," a term with a powerful meaning: We can no longer rely on past events to predict future probabilities.[1] The ground is shifting beneath our feet.

Now it's becoming increasingly clear that the volatility we're seeing in our natural world is reshaping our financial world, too. While we try to hold onto our bearings in the status quo and conduct business as usual, new risks are entering into the equation, and new costs are creeping onto the balance sheets of corporations and nations. Just as chaotically as climate change is altering conditions here on earth, it is changing our understanding of the actual costs of just about everything.

In the financial realm, as in the natural realm, the past provides fewer and fewer clues to our future. Like the migration patterns of songbirds that no longer correlate to the hatching patterns of their insect prey, or the mountain snowpacks that no longer store water for the dry summer months, the economy is facing similar miscues borne of the interactive loop between tumult in the atmosphere and tumult on the earth.

Economic calculations are being upended in ways that are similar to how the digital revolution transformed financial decision making—though on a far more monumental scale. In early 2014, a report by the Intergovernmental Panel on Climate Change (IPCC), produced and reviewed by more than a thousand scientists and approved by 194 governments, identified the economic impact of the enormous changes under way: slowing the rate of greenhouse gases into the atmosphere would not have devastating impacts on the economy, as is often feared, but would require a fundamental rethinking of our priorities and our existing business models. Unstated but clearly implied: new forces will rise and others will fall as the true costs of carbon become an increasingly potent economic factor that no longer can be ignored. In 2013, an insurance industry research association called for an entirely new paradigm for assessing risk, because rapid-fire changes in weather and temperature are outpacing traditional actuarial calculations.[2]

In other words, we are on the threshold of carbon shock.

There are five greenhouse gases. Together, these gases, of related though different provenance, are in a circuit of relationships that give them the same ultimate impact—trapping heat in the earth's lower atmosphere, giving rise to the greenhouse effect. Up from the center of the earth comes the hundred-million-year-old periodic table of decaying organic fossils in the form of CO_2, carbon dioxide, and CH_4, methane, embedded in oil, coal, and natural gas—thus the term *fossil fuels*. From the combustion of those ancient fossils comes O_3, otherwise known as ozone, derived from the interaction of sunlight with CO_2 and other hydrocarbons; NO_2, nitrous oxide, which in addition to being emitted from industrial sources is also contained in fertilizers and released from soil through agricultural tilling; and a synthesized family of hydrofluorocarbons, generated as waste products from the production of other chemicals, primarily refrigerants and pesticides. In the gas world, they are all cousins, baton passers of molecular reaction that conspire to create turmoil in the heavens. Carbon dioxide is by far the most abundant, which is why *carbon* has become a stand-in reference for all greenhouse gases.

It's not that carbon hasn't always had a cost; it's just that, until now, carbon's costs have been, for the most part, invisible. We haven't been accounting for them. Just as an optical illusion tricks the eye into seeing something that's not there, traditional accounting diverts our attention from invisible costs; we see only profits. Keeping these costs a mystery has been fundamental to our economic growth. Businesses benefit from a false ledger in which the environmentally corrosive impacts of the energy they use to produce, distribute, and dispose of everything from almonds to automobiles have been dramatically undercounted. The world's three thousand biggest companies, according to a United Nations Environment Program report, cause $2.15 trillion in annual environmental costs.[3]

Most of those costs are the result of greenhouse gases emitted from the burning of fossil fuels, from which 82 percent of the world's energy is derived.[4] Unlike smog or smoke or other acute contaminants, you can't necessarily see greenhouse gas pollution. But we are the involuntary recipients of its steadily expanding effects. These effects come to us fractured, in scattered reports of distress: polar bears floating off on melting glaciers, hurricanes tearing into distant coastlines, droughts sending farmers into panic, remote island states sinking into the sea.

When in the summer of 2013, a gas chromatograph perched atop the Mauna Loa volcano in Hawaii reported that the concentration of greenhouse gases in the atmosphere had for the first time reached four hundred parts per million, it not only triggered alarm bells from scientists, but signaled costs to come. The world's two biggest economies, the United States and the European Union, estimate hundreds of billions of dollars in costs from heat waves, floods, and an accelerating flow of refugees fleeing lands in which they can no longer sustain themselves. Both classify climate change as one of the foremost challenges to political stability. The World Economic Forum has identified erratic water supplies as one of the primary challenges to economic stability. The Food and Agriculture Organization predicted rising food prices as conditions shift toward a perfect storm—lower rainfall in already dry areas, and more torrential rainfall in areas that are already wet. Climate change, in short, is the most significant economic challenge of our time.

Somebody does pay for those costs. That somebody is us. This concept is so deeply embedded that economists have a name for it, externalized costs—the costs borne not by the producer or the immediate consumer, but by society.

In the United States, the government has assigned an average figure of thirty-eight dollars per ton of greenhouse gases for what it calls the "social costs" of carbon[5]—which might as well be called the externalized costs of carbon. This figure is supposed to account for the costs to the nation's preparation for and recovery from extreme weather events; the pressure on coastal infrastructure from the rising sea; the public health impacts as diseases, pests, and bacteria once limited to the tropics move northward; the rippling effects of ever more erratic rainfall; and the loss of productive capacity from all these phenomena. In 2012, the United States was responsible for 6.5 billion tons of greenhouse gas emissions,[6] putting a significant price tag on American emissions. Such figures, though, are calculated according to political as well as economic factors; they can vary widely depending on how broadly the impacts of climate change are defined and the rate at which future damages are valued. The UK government, for example, assigns a figure of fifty-five to eighty-three dollars per ton for the social costs of carbon.

We're now living in an era being transformed physically and financially by those externalized costs. Over the past fifty years, concluded the 2013 National Climate Assessment, there have been an unprecedented number of heat waves, severe droughts, and heavy rains, climatic changes that it said are "primarily due to human activities." Recovery efforts from the hurricanes of 2012—the severity of which was attributed at least partially to climate change—amounted to forty billion dollars. And the costs only accelerate as long as emissions continue at the current rate: We use more energy to cool ourselves as the weather gets hotter; farmers import water from ever-more-distant locales at ever-increasing prices; levees are reinforced or in some instances demolished to make way for rising seawater.

The first person to estimate in detail the financial burden created by climate change on a planetary scale was an economist at the London School of Economics, Nicholas Stern, who reported to the British government in 2006 that it would require the equivalent of at least 1 to 2 percent of global GDP, by 2050, to adapt to and respond to those costs; seven years later, as more data rolled in, Stern revised that figure upward to 3 to 4 percent of global GDP—more than a trillion dollars by the end of the century. By 2014, he was saying that measuring climate change in terms of GDP is too limiting, and does not take into account the massive displacement and loss of life from floods and droughts if current emission trajectories continue.

However you count, it's a lot of money, enormous costs that are uncounted by those who create them. They're off the official books because we, the public, pay for them. The emitters of greenhouse gases get a free ride. And we do, too, subsidizing ourselves up front with cheap energy—good for the profits of the energy companies—only to pay far more later when the costs come due. This is known as asymmetric risk, a term of the financial arts that means the public bears the risks while fossil fuel users earn the profits. What it means on the ground here where we all live is that we've been engaged in an immense transfer of wealth and resources from our global commons.

When the largest US oil companies agreed to calculate a price for greenhouse gases in their long-term investment scenarios in late 2013, this signified a partial reckoning with the symmetry of risk that's been a long time coming and sent shock waves into the system. Now, though, there's no turning back. Like other efforts to rebalance the books, it lifts the lid on the trick behind our accounting magic: the actual cost of energy that sleight of hand has long kept out of sight.

———

Fifty billion tons of greenhouse gases were emitted worldwide in 2012—a figure with so many zeroes it is difficult to fathom. During the climate negotiations in Copenhagen in 2009, the Danish Energy Agency helpfully installed a huge yellow balloon over the city's main square that stated in bold black letters etched into the outline of a globe: THIS IS THE SIZE OF ONE TONNE CO$_2$. The installation, billowing in the icy breezes, was enormous—about two stories high and a block wide, the size of a hot-air balloon that could carry you into the atmosphere. Its presence loomed as passersby hurried underneath it during that snowy December. So you would need fifty billion of those balloons—filled with the gases generated primarily by utilities and oil refineries, coal-powered manufacturing, transport, agriculture, and decaying and dying trees—to see what the greenhouse gas threat actually looks like. The balloons hang invisible above our heads, altering the atmospheric balance and thus the balancing act of life here on earth. And every year there are more balloons, because climate change is cumulative: The gases just keep piling up as long as we release them. John Harte, a climate scientist at UC Berkeley, prefers the term *overcoat* to *greenhouse effect*, because the layers of atmospheric insulation keep getting thicker as the gases accumulate.

Those balloons full of CO$_2$ might as well contain cash, depleted from the world's coffers with each new ton. Up go balloons full of money. Seeing climate change through the prism of its costs may be the only way to overcome the illusions that have constrained us from making an honest assessment of our options. Fossil-fuel-based production is favored through misleading accounting, coupled with massive subsidies. According to the International Energy Agency, the subsidies given to fossil-fuel industries amount to more than five hundred billion dollars a year worldwide,[7] contributing to the misleading calculation that fossil fuels are the most economically viable form of energy. Using public funds to subsidize the fuel source that is undermining conditions of life on earth defies logic, not to mention the principles of the free market.

Shifting our economic engines in a more renewable and resilient direction requires examining the very means that humankind has used to power itself over the past two centuries. It's as if a shock to the muscular system of the human body suddenly altered how we move, so that our gait, our mobility, could no longer be taken for granted. Climate change is that shock to the global economic order. We need to find new ways to move.

But what is a carbon footprint? And what is our contribution to that footprint? What are the costs of climate change? Who has the responsibility to pay for these costs? Whom do we pay?

The attempt to answer those questions sent me on a journey to places where new economic angles are emerging and the axis of geopolitical power is shifting. Climate change is the single most effective eye-opener to how globally connected we all are—from tropical rain forests to oil-splattered beaches, from designers of cities to farmers struggling with the changing elements to traders betting on the price of carbon, the corrosive effects of climate change unite us across national frontiers, as does the fight to slow the change down. And as I would also discover, it is the elusive cost of carbon, paid out and fought over in so many different forms, including as one of the world's newest and most bizarre financial commodities, that is the greatest economic disrupter of the twenty-first century. There is unrest in the corridors of financial power, where it's becoming clear that the "end of stationarity" is as applicable to our understanding of the earth as it is to our understanding of the new risks for economic decision making.

When it comes to risk, our brains are generally wired to see those right in front of us; we perceive patterns of threat that spur the fight-or-flight instinct. But the threat from climate change is of a different order—occurring in different forms, some dramatic and some subtle, all over the earth simultaneously. "Our risk management patterns are still wired to search for lions in the Serengeti," commented Mark Trexler, CEO of The Climatographers, a consulting firm advising businesses on how to adapt to environmental risks. "See lion—run. That's what we're still doing in the climate space."

But lions in the desert are not the threat. The Serengeti itself, that seat of human life that is a stand-in for the planet, is being transformed. The patterns we are now living through have never been seen before—they are the patterns of departing from previous patterns.

Into this world in flux we will travel in search of the price of carbon. Each of us contributes to blowing some portion of those gases into those balloons filled mostly by emissions from the United States, China, and Europe. We'll follow a trail that will take us deep into the backstory of our carbon footprints, largely hidden from public view, and go behind the scenes into life's most common activities: we travel; we drive; we eat; we buy goods that make our lives more comfortable and amusing; we inhale the same oxygen that comes from trees, which inhale carbon dioxide, and release it when they're burned or cut down.

Unlike other such trails, though, ours won't lead necessarily to who's spending money on which politician or interest group; that has been done, to great effect, by others (and most of those trails lead—surprise!—to oil companies and fossil-fuel-reliant manufacturers). Instead, we will travel into an ever-widening circle of people and places where the costs of climate change are already being acutely felt. Our journey will land us in the midst of some of the most powerful interests of our time—some of which are trying to move forward while others are trying to block the transformation to a twenty-first-century economy that delivers far less collateral damage. All are maneuvering for position while climate change shifts the conditions around them—and while the risks from the status quo grow.

One way to think about carbon footprints is as the embodiment of financial risk. If companies do not yet see the risks, their insurers certainly do. "We foresee," concluded Lloyd's of London, the scion of global insurance companies, "an increasing possibility of attributing weather-related losses to man-made climate change factors. This opens the possibility of courts assigning liability and compensation for claims of damage." Add to that the potential disruption of production and supply chains; the reputational consequences of consumers and investors becoming more aware of the environmental underside of their favorite products; and regulatory moves by governments, which are fitfully but increasingly instituting penalties on the emission of greenhouse gases—and the risks mount. Few of these looming triggers, however, are required to be reported to potential investors—though any one of them could seriously undermine the financial value of companies reliant on fossil fuels.

For almost two decades, negotiators have attempted to redress the imbalance between who creates the risk and who pays for it, to forge a price for carbon that reflects the vast differences in responsibility for our situation and that is steep enough to trigger a shift away from greenhouse-gas-producing energy. Cap and trade—a system by which major greenhouse gas emitters were given emission caps, and could buy, or "trade," allowances if they emitted more than their allotment—was the first effort to try jump-starting a new way to value pollution. Polluters would pay for their contribution to climate change up front by subjecting carbon to the supply-and-demand forces of the market. This was a way "of assigning monetary value to the earth's shared atmosphere," declared the United Nations, "something that has been missing up to now."

First proposed by the United States in 1997 during the negotiations in Kyoto, Japan, this was also the preferred option of industries that perceived it as the least expensive method for dealing with their greenhouse gases. Then it was cut loose back in 2001 when the Bush administration withdrew the United States from ratification of the Kyoto Protocol. "Kyoto is dead!" famously proclaimed the then national security adviser Condoleezza Rice.

But Kyoto did not die. Europeans were left to implement a program designed largely by Americans. The strategy has had a troubled history, but one thing it clearly accomplished was to create a new set of global fault lines between countries that have at least a minimal price for carbon and those that do not and which are subject to few carbon constraints. So instead of one price, which had been the intent in Kyoto, we've got wild variation and huge rifts—divisions that hinder the search for global climate solutions.

"We are trying to prevent murder in the future" was how M. C. van Leeuwen, an agent with the Organisation of the Dutch National Police, described to me his job fighting environmental crimes when we met at a conference of environmental police sponsored by INTER-POL. He was taking poetic liberty with the term *murder*, referring to the difference between investigating a traditional homicide and the evidentiary trail behind the slow, steady wounding of ourselves that results from humankind's erosion of life-sustaining forces on the planet. Van Leeuwen saw environmental crimes as their own form of murder, of the web of living and natural forces that give us a hold on the future.

Tracing back the fingerprints, we know who's behind them. Just ninety companies, according to a peer-reviewed study in the journal *Climatic Change*,[8] are responsible for two-thirds of greenhouse gas emissions.

My approach, in the pages ahead, is that of a detective on the trail of the financial crime that lies behind those emissions, and that is transforming our world, a theft from all of us. It's a crime that's perfectly legal (for the most part), because the rules have been written to accommodate an illusion that has served our economic development well. But the illusion is in tatters. The costs—payback for our energy discount—are bursting out all over.

Meanwhile, a new economy emerges amid all these competing forces. Indeed, a boom is under way. Had it not been for the sporadic way it has taken shape, this "green" economy—constituting the hundreds of billions of dollars invested in renewable energy innovations and emission-reduction technologies in the United States, Europe, and, increasingly, China—would have long ago been considered game changing and disruptive, akin to how the tech bubble transformed investment decisions and shook up the landscape of economic power. Now it exists in haphazard pockets. But they are the leading edge of the transformation under way.

The two key premises on which this book is based are that the costs of climate change are occurring now, not in some distant and abstract time in the future, and that identifying the actual costs and consequences of fossil fuels will give us a far more financially honest way to face our premier economic challenge. It will give us a foothold into the shifting landscape at a time when predictability is hard to come by and the ability to prosper requires being able to read the signs coming at us from every corner of the earth.

I set off to find some of the key tension points in this evolving new economy, places where we can see both the impacts of climate change being experienced today and the struggle to create a new economic order that will shift our current trajectory.

Our journey begins in the air above the North Atlantic.

CHAPTER I

Dogfight Over My Flight

Airplanes

*I*t was a fine, clear day in the atmosphere as I flew from San Francisco to Siberia. Through the window, below the wispy clouds, I could see through to the icy flatlands of the northern latitudes en route to our first stop, in Paris.

The long flight was typical—a few bumps over the Atlantic, bad food, bad movie, plenty of time to sink into my pile of magazines and my favorite Russian author, Mikhail Bulgakov. Our Boeing 747 was just one more airplane flying over the footsteps of humankind.

Across North America we flew, over the fields that are faced with new stresses as climate change alters the growing conditions for our farms. Over the cities, where most of us live, that are grappling with how to reconstitute themselves in greener directions. Over the land that once grew forests now known to be one of our most effective means to soak up carbon emissions. We passed over the western coast of Spain, which has experienced its own tragedy from our obsession with oil. And we veered just to the south of London, center of the world's effort to subject carbon to the forces of the market. Behind and in front of us were the countries, Brazil and China, that face the challenge of growing quickly at a time when carbon, the offspring of all those fossil fuels we in the United States have obtained at such a discount, is for the first time taking on a price.

These are the places where we land throughout the pages of this book. If I'd been in the space station orbiting two hundred miles above my plane, I might have caught a more impressionistic view of them—the marvel of our mountains, our forests, our oceans, all discernible in patterns of blues, greens, and browns to the astronauts given that privilege. But of course I was a mere eight miles high, and eventually I landed peaceably enough on the continent that has been engaged in a monumental though barely noticed fight with the United States over the emissions coming from the back of my airplane.

Down here on earth, far below the public radar, the dispute was brewing: There were whispers of a global trade war, US and international carriers were threatening to boycott European air routes, and US airlines were pursuing the first-ever legal challenge by an American industry against the EU.

The conflict was causing more turbulence on the ground than we were experiencing in our five-hundred-mile-per-hour passenger jet hurtling through the atmosphere eastward. As I flew, judges at the European Court of Justice, charged with arbitrating disputes involving the European Union, were considering their verdict. The world's two biggest economies, the United States and Europe, were fighting over whether to account for the greenhouse gases coming from my plane that day.

My plane, of course, was not the only target. I was one of thousands of passengers flying from the United States to Europe that day in one of hundreds of planes traversing the air routes of the planet. We were stacked in the sky like a mobile seven-layer cake. All of those planes, as they are each day, are the threads that connect the distant outposts of our global community—and emit greenhouse gases every inch of the way.

My own contribution that day, according to the International Civil Aviation Organization, was fifteen hundred pounds of CO_2, simply by virtue of being a passenger on a fossil-fueled airplane[1]—a minuscule contribution to the larger picture of aviation emissions.

Aviation contributes 3 to 4 percent of the greenhouse gas load every year, according to the United Nations, more than any other form of transport. Electrical utilities are the only single industry with a larger carbon footprint. World air travel has doubled over the past twenty years, and it's expected to increase exponentially: The Congressional Research Service said that aviation is "one of the fastest growing sources of CO_2 emissions."[2] By 2020, the emissions coming from airplanes leaving from or coming to the United States alone are expected to be 75 percent higher than what they were in 1990.[3] "If you want global warming, then traveling by air is one of the fastest ways to get there," said Bill Hemmings, policy director for Transport & Environment, an environmental NGO based in Brussels.

Recent science suggests that because of the altitude at which those emissions occur—high enough to be injected directly into the atmosphere—they are likely having a more destructive impact than similar emissions would have on the ground. That's because the nitrogen oxides and water vapors emitted by airplanes create "contrail clouds" high up in the atmosphere, which further intensify the insulating greenhouse effect. The higher you go, the more destructive the impact from the wastes of burned jet fuel.[4]

Of course, one of the tricky things about airplanes is that their pollution comes from everywhere and nowhere. Planes in the air make no distinction about national borders. It's not as if the emissions from an airplane taking off from San Francisco just hang around over California, or those from an airplane taking off from Beijing remain over China. Yet no matter where they're from, greenhouse gases have the same disruptive impacts on the earth. Aviation provides a window into the global fault lines created by climate change. While most Americans and Europeans will at some point travel on an airplane, churning out their greenhouse gases with every mile, the majority of the world's population, who will be equally if not more acutely impacted by the effects of those gases here on earth, will never fly in a plane, or will do so far less than we do.

Each metric ton of burnt jet fuel produces three tons of CO_2,[5] bringing with it a bundle of financially uncounted consequences. Yet there are no taxes levied on aviation fuel to account, even minimally, for its environmental impact; the roughly 20-cents-per-gallon jet-fuel tax goes to support the regulatory and safety functions of the Federal Aviation Administration. You'd pay more in environmental gas taxes driving five miles to a lunch in Los Angeles than you would flying from California to Russia, as I did. Two analysts at the International Monetary Fund, bastion of realpolitik economics, concluded that there was a "strong" case for more taxes on aviation. "[T]he indirect tax burden on international aviation is very low," they wrote in the economics journal *Fiscal Studies*, "yet aviation contributes significantly to border-crossing environmental damage."[6] My contribution to the disruption of the atmosphere constituted a free ride.

But someone, somewhere, at some point, would end up paying for the environmental costs of my flight. One thing the airlines wanted to ensure was that it wouldn't be them, or me, at least not while I was flying on one of their planes.

Where does national sovereignty begin and end when trying to limit the flow of greenhouse gases into our shared atmosphere? Whom, if anyone, would I pay for my airborne carbon footprint? As I hurtled across the stratosphere, these were the questions over which a historic battle would be fought.

It began with a relatively modest effort by the EU to attach a price to the greenhouse gases coming from airplanes. Just a few months later, in 2012, the EU began requiring airlines to obtain allowances for their emissions en route to or from European airports. The aim was to impose a cost that more closely resembles the actual costs of airplanes' contribution to climate change—and to achieve a 3 percent drop in airplane emission levels from 2005, a rate of reduction that increases in progressively greater amounts over the coming decades. The EU estimated that the scheme would eliminate seventy million tons of CO_2 annually, roughly the yearly output of sixty thousand automobiles. It would generate (assuming a carbon price of ten dollars a ton) from three to eight hundred million dollars a year in funds that the EU suggested be used for climate mitigation in Europe and around the world. Eighty-five percent of the allowances were to be issued for free in the first five years. That percent would decrease over subsequent years, and the remainder would be purchased on the carbon markets—intended to set escalating prices for aviation-based greenhouse gases.

"What we're trying to do is send a signal," said Philip Good, who was charged with implementing the new rules at the Transport Directorate in Brussels. "If people emit carbon [in planes], they should pay for climate mitigation around the world." It would also, he said, spur investments to put Europe at the forefront of developing less fossil-fuel-intensive sources of power for jet engines.

In addition to the twenty-eight member states of the EU, Norway and Iceland signed on. All discovered that they were stepping onto the trip wires left behind by negotiators' failure to devise a new global climate treaty. They would soon be thirty countries against the world, engaged in a struggle that would reveal just how much our twenty-first-century understanding of climate change is reshaping twentieth-century concepts of national sovereignty and international law.

It would also shuffle the deck of geopolitical alliances, as the United States found itself in league with a host of unlikely allies opposed to the EU, including China, India, and Russia. Flying at thirty-five thousand feet through the airspace of a dozen countries, I had become a bit player in a global conflict over the cost of my flight to the future of the planet.

The first move of the US aviation industry was to try an unprecedented legal attack. Three US air carriers—United, American, and Continental (later merged with United)—lodged a challenge to the law in a British court, which ruled that the case should be handled by the EU's court. The dispute traveled to the European Court of Justice in Luxembourg, then steadily escalated until reaching the United Nations, the US Congress, and finally the doorstep of the earth's fastest-growing economies, where air traffic is expected to explode as a source of emissions over the coming decades.

The undisputed fact of the conflict was in the emissions contribution of airplanes. Over the course of my fifty-eight-hundred-mile flight from San Francisco to Novosibirsk, with stops in Paris and Moscow, 330,000 pounds of greenhouse gases were emitted from my plane according to Air France, my carrier. Like every other flight reporting emissions, its impact was a known quantity. Had we been flying to, say, New York, no one would have counted the CO_2 and other greenhouse gases emitted from our Boeing 747. But we flew over the United States, and by the time we landed at our first stop in Paris, the numbers had been lodged with the European authorities.

Midway through our flight over the Atlantic, I asked a steward if I could have a chat with our pilot—to whom I'd introduced myself shortly before takeoff. An hour later, as our plane cruised through calm air, Bertrand Villepique strolled down the aisle from the front of the plane—drawn outside the cockpit, he said, by the novelty of my request to talk about our plane's greenhouse gases. Standing between the food locker and the toilet, we chatted for a few minutes, and Villepique told me that the greenhouse gas emissions from our flight, and all others between France and the United States, were reported cumulatively every six months to the French Ministry of Transport. The airline did the same for its many flights inside Europe and to other parts of the world. It was routine, he said, "just another number" on a report to headquarters, like how much gasoline the plane used or how many meals were packed into its refrigerators.

Villepique had to submit no such comparable data to the US Federal Aviation Administration. FAA Public Affairs officer Hank Price told me that the agency strives to reduce US carriers' fuel consumption by designing less fuel-intensive routes, but does not correlate that data with greenhouse gas emissions. While Air France's US partner airline, Delta, awarded me "miles" for my trip, it was under no obligation to report its emissions to anyone unless it landed in or launched a plane from a European airport. Indeed, US–European carrier alliances, like Delta's SkyTeam partners Air France/KLM, were poised against each other in this conflict. In order to create a level playing field, the major European carriers insisted that their US counterparts be subjected to the EU law, which US airlines claimed did not apply to them.

I changed to another Air France plane in Paris, and onward we flew—emitting another half ton of CO_2 en route to Moscow that was reported to the French Transport Ministry. In Moscow, I caught an Aeroflot plane and flew onto Novosibirsk, the heart of the Siberian economy, home to major resource and oil companies and one of Russia's leading universities. Sophisticated studies by Russian scientists have rendered detailed data on the contributions to climate change made by the country's air carriers.[7] But just as in the United States, no one in the government was counting. I spent two weeks in Novosibirsk and elsewhere in Russia conducting trainings of some of the country's top environmental journalists, not one of whom (like most of their American counterparts) had the slightest idea that their national government was overcoming a decade of tension with the United States to become one of its most aggressive allies on the hot-button issue of holding airlines and passengers accountable for their contribution to the costs of climate change.

The single disputed fact over what everyone conceded were greenhouse gases being emitted by my plane and others is whether we would have to pay for the damages. Played out in diplomatic corridors and courthouses, the dispute over this fact would become the world's first battle over how to assign a price to a singular carbon footprint. The world had long ago gathered in the Japanese city of Kyoto to devise a global plan to give a price to carbon. Now, some fifteen years later, the battle had advanced to assessing how to penalize the carbon footprint for an individual airplane and, in essence, an individual passenger—me.

The scene, July 2011, 9 AM: In the sleepy principality of Luxembourg, inside the modernist glass-and-steel chambers of the European Court of Justice, a panel of thirteen judges in dark red robes considers whether the EU has the legal right to penalize the carbon footprint of an airplane. Facing them from a table to their right are lawyers representing United, American, and Continental airlines, along with their trade group, the Air Transport Association (ATA, since renamed Airlines for America). To their left sit lawyers representing the European Union, its member states, and a lawyer representing five environmental organizations, including the Environmental Defense Fund, World Wildlife Fund–Europe, and the Brussels-based Transport & Environment, all of whom the EU had welcomed—an unprecedented development in itself—onto their defense team. In the front rows of the observer gallery sit officials from the US State Department and from the aviation agencies of the United States, China, and India.

Here the price of our mobile carbon footprints in the air began to take real shape. The array of some twenty lawyers, high-powered government representatives, aviation trade groups, and international environmental NGOs suggests the broad implications of a case that nevertheless received little attention in either the US or European press.[8]

Back in San Francisco, as in many other airports, there were automated kiosks where I could purchase an "offset" to my flying carbon footprint. But these are subject to no legal verification as to whether your good-conscience investment actually delivers the carbon sequestration or emission reductions promised. The EU was upping the stakes significantly by making mandatory what had been purely voluntary—for airline passengers, at least, if not carriers—and putting legal teeth behind more accurate fuel accounting.

The EU was on new legal territory here, edging its way into our atmospheric commons. Creating legal mechanisms for dealing with the earth's other common resource—the oceans—had long ago been settled. The Law of the Sea specifies sovereign rights over a nation's territorial waters, which means that countries can impose their own safety standards on what kinds of oil tankers, for example, are permitted to dock in a nation's ports or pass through its coastal waters—a measure prompted by the *Exxon Valdez* oil spill in 1989. Could similar principles, drawn from the threat of greenhouse gases—a sort of slow-motion oil spill in the heavens—be applied to the infinite sky? For six hours, the question before the court got down to the brass tacks of who, if anyone, has the right to regulate the greenhouse gases emitted by border-crossing airplanes.

The airlines' chief lawyer, a British barrister named Derrick Wyatt, centered his legal assault on defending the traditions of national sovereignty. He conjured a scenario very much like my own. In a flight from San Francisco to London, he told the court, 29 percent of the emissions take place in US airspace; 37 percent in Canadian airspace; another 25 percent over the high seas; and just 9 percent in the airspace of Europe. The measure, he said, represented an unprecedented extension of European authority, amounting to "EU regulation of US airlines in third country airspace and over the high seas." Further, he contended, the proper place to administer greenhouse gas regulations on airplanes was not the EU, but the International Civil Aviation Organization (ICAO), a 180-member UN group that sets broad guidelines for the world's air traffic.[9]

Unstated, however, was the fact that for more than a decade, the American aviation industry and successive US administrations had opposed efforts to do precisely that. First, after intense lobbying by the airlines, aviation was excluded from the emission mandates of the 1997 Kyoto Protocol—which instead placed the responsibility with ICAO. The EU made repeated proposals to ICAO to begin regulating airline emissions; the Bush administration blocked every one of them. By 2004, the agency itself announced that it did not have the legal teeth to devise a system for global emission controls, and shifted responsibility back to individual countries. The EU took them up on the offer. Then Bush's aviation team undermined even that proposition when they pushed a resolution through ICAO that any country regulating greenhouse gas limits on international flights could only do so with the "mutual consent" of the affected parties—which meant that each country had to agree, making it in effect impossible to create a system for governing international flight emissions.[10] The EU filed a formal "reservation" with the agency, and said it would not comply. That history was missing from the airlines' presentation.

After a decade of trying for a global approach, the Europeans moved forward on their own. "Over and over, the US blocked efforts on emission limits at the International Civil Aviation Organization," recalled Annie Petsonk, who represented the Environmental Defense Fund at the court hearing. "And they did essentially nothing. The EU got fed up with getting nowhere and said, 'Okay, we're going to start to regulate.'" That's why Mr. Villepique, the pilot of my Air France flight, was reporting our plane's emissions. For the first time, a price was lodged on the carbon footprints of air carriers.

Europe's attorneys, forging into a new international legal frontier, stuck closely to the facts of the case. They insisted that Europe was not proposing to regulate what airlines do in international airspace, but rather was demanding that the airlines deal with climate change if they want to utilize a European airport.[11] And they claimed that the principle of extraterritoriality in aviation was in fact initially forged by the United States—which in the wake of the terrorist attacks on September 11, 2001, demanded that the security requirements for airplanes bound for the US follow American guidelines. That's why foreign airport security mirrors almost precisely that of airport procedures inside the US. Just as the United States determined that the threat of terrorists after 9/11 warranted global security precautions, the EU argued that the urgency of climate change required a similarly global response.

Peter Liese, a member of the European Parliament and the law's chief author, described the measure as "our attempt to start to compensate for the externalized costs of aviation." Liese is a German Christian-Democrat, on the conservative side of the European political spectrum. He told me that the issue was not ideological, but economic: The Parliament wanted airlines to compensate for the actual environmental costs of air flight. While the new law had the support of environmentally oriented parties like the Greens and the Socialists, Liese's Christian-Democrats, known for their business-oriented tilt in Germany, saw it clearly as an accounting question—reimbursing the government for at least some of the billions of dollars in costs to Europe and to the planet from airplanes' contribution to climate change.

The law was endorsed by major European and American environmental organizations—which hoped it would kick-start the development of less atmospherically destructive jet fuel. "This represents an opportunity for US companies to compete and succeed in a world where carbon constraints are real," commented Pamela Campos, a senior attorney on the case with the Environmental Defense Fund. "We've been telling the administration, 'This is your opportunity to kick butt, take the lead and drive our innovation guys to do the innovations and push Europe and the rest of the world.'"

But Obama's position was unchanged from that of his predecessor. At ICAO meetings, US officials continued to block European efforts to control global emissions. One of the key negotiators for the United States at those meetings was Julie Oettinger, the FAA's vice director for international affairs—who until 2010 had been United Airlines' chief lobbyist on Capitol Hill. This put her in the position of arguing and indeed helping to design a US position that was, at the least, in parallel with that of her former employer and, at most, destined to benefit the airline (at least on a short-term basis) if the US could successfully dodge responsibility for airlines' greenhouse gas emissions.

Emily Cain, spokesperson for the State Department, told me, "The [US] government has made it clear to the EU that we do not agree on both legal and political grounds with their unilateral plans to impose policies on other countries."

Back in Luxembourg, five months after the airlines argued their case at the European Court of Justice, the judges issued their verdict: The EU's actions, they ruled four days before Christmas in 2011, were consistent with international trade constraints and with European law.[12] Charging for our aviation carbon footprints was given a green light.

Diplomatic warfare commenced shortly thereafter. In 2012, the transatlantic air routes became the fault lines in the first global trade war over the cost of a carbon footprint.

Days after the court's verdict, Secretary of State Hillary Clinton and Secretary of Transportation Ray LaHood sent a joint letter to EU president Manuel Barroso demanding that the EU "repeal or postpone implementation" of the aviation directive. Two months later, in February 2012, the United States mustered Indian and Russian aviation authorities to a meeting in Moscow, where twenty-three countries signed a declaration echoing their letter—demanding that the EU "cease application" of its directive to "airlines and aircraft operators registered in third countries."[13] Even in the annals of cantankerous global jousting over climate change, the attempt to intervene with the already existing law of a democratic government was unprecedented. The equivalent would be asking an American president to repeal a law he'd signed after it was passed by Congress. Signatories to the Moscow Declaration also threatened a litany of countermeasures—triggering a diplomatic dogfight over our carbon footprints in the air.

Russia announced that it would close Russian airspace to European planes flying over its territory. Chinese and Indian aviation authorities forbade their nation's carriers from complying with the EU's demand for greenhouse gas data. China exerted its economic heft by pulling out of a multibillion-dollar deal to purchase several dozen new Airbus planes for its rapidly growing fleet.

For those and other developing countries, the EU's measure threatened a treasured principle of what's known in climate circles as "differentiated responsibility"—the distinction between a developed country's and a developing country's responsibility for climate change. The EU's aviation scheme would treat both equally—a fundamental advance from previous greenhouse gas agreements that distinguished Chinese and Indian emissions from those of, say, the United States, Europe, and Australia. They particularly did not want that principle challenged in advance of the next round of renegotiations of the Kyoto Protocol in 2015.

Two weeks after his reelection, President Obama aligned the United States even more closely with India and China when he signed a law, another precedent breaker, that would allow the secretary of transportation to forbid US carriers from complying with the European greenhouse gas requirements. The law, authored by Republican senator John Thune, had the potential to force US airlines into an impossible dilemma, a choice between defying American or European law. Transatlantic traffic could be brought to a standstill. Altogether, countries representing more than half of the world's air traffic refused to comply with the EU. The effort to contain our carbon footprints in the air had certainly hit a nerve.

One interesting result of this conflict was that it ultimately triggered the release of airline emission data never seen before. The Federal Aviation Administration requested that the EU give it the greenhouse gas records that international airlines had given to the European authorities—those very same records submitted by my Air France pilot. The FAA's intention, though, was not to determine the quantities of greenhouse gases that were being released; rather, it was intended to send a signal that the United States was preparing to take retaliatory action at the WTO or other international trade bodies. "The Department of Transportation is now at the frontlines of a trade war with Europe," commented Gabriel Sanchez, a fellow at the International Aviation Law Institute at DePaul University in Chicago, at the time.

Some eighty carriers from around the world, including those from the United States, submitted data to Europe's aviation ministries, which they then turned over to the FAA. Suddenly the FAA was in possession of airline-by-airline details of greenhouse gas emissions that it does not have the power to obtain in the United States.

I, too, obtained a copy of the data. Pages and pages of numbers across spreadsheets told the story: From Albania to Zimbabwe, the world's airlines reported their greenhouse gases for 2011. This was information each airline had turned over to national transport authorities in countries where carriers have their hubs—for example, United reported to the British Aviation Ministry, Delta to the German equivalent, because of their respective hubs in London and Frankfurt. Each ton of emissions they reported then had to be covered by an allowance submitted in the following year. This was the first time that a comprehensive worldwide tally of an airline's greenhouse gases had ever been assembled. But while the FAA scrutinized the numbers for evidence of perceived discrimination against US carriers, to me this represented a data gold mine. The numbers revealed an avalanche of emissions, airline per airline, remarkable in its specificity.

There they were, the plaintiffs in that case at the European Court of Justice. American Airlines came in at 2,745,318 metric tons of greenhouse gases emitted by its planes going in and out of Europe. Continental (now merged with United) came in at 2,146,690 metric tons; United, 2,440,010 metric tons. Freight carriers also fall under the EU's law: FedEx, which has a major transit depot outside of Paris, emitted 1,514,944 metric tons while shuttling between the United States and Europe; and UPS (United Parcel Service), with a major depot in Germany, emitted another 1,307,002 metric tons. Even Starbucks gets a ranking, registering 26,000 tons (presumably for its private corporate jets and coffee-freighted cargo planes), along with McDonald's, which came in at 27,000 tons (presumably for its planes loaded with meat, not to mention company executives). US-registered airplanes were by far the biggest non-European greenhouse gas polluters; European carriers must account for their emissions inside Europe in addition to international flights.

Notably, of course, the dozens of US carriers that do not fly to Europe were not on the list, so there's no way to know the emissions of, for example, Southwest Air, Jet Blue, Alaska Air, or any other of the regional airlines of the United States. Or, for that matter, the emissions of inter-US flights, or flights to anywhere other than Europe from those that do actually fly to Europe. Among the Europeans, British Airways was the top emitter, followed by Lufthansa, Air France/KLM, Virgin Atlantic, and Ryanair (a discount carrier that makes countless short hops across the continent). Oh, and there was the Russian Aeroflot, coming in at 626,440 metric tons.

Numbers, metric tons, lines upon lines, each a strand of evidence about how an airplane's flying time translates into erosion of the global atmosphere. All those numbers made the outlines of the battle starkly clear: If forced to account for their greenhouse gases, air carriers from almost every country would have to upgrade the price of a ticket.

The actual cost was disputed wildly. The Air Transport Association, party to the unsuccessful EU lawsuit, predicted jumps of as much as forty-five dollars per transatlantic ticket. The EU asserted that the cost would be more like six to ten dollars per ticket. A study by MIT came in with a conclusion that should have, but did not, dampen hysteria on both sides. The European initiative, MIT concluded, "would only have a small [financial] impact on US airlines and emissions."[14]

By the fall of 2012, a fuel surcharge was quietly slipped into the price of a round-trip transatlantic ticket by the three airlines that had sued the EU. The airlines—Delta, American, and United Continental—refused to publicly identify the charge as covering the cost of their greenhouse gases. Rather, it appeared quietly without mention just as they were faced with the prospect of purchasing their first round of emission allowances. The cost added to those tickets came to a hardly devastating six dollars. Over that same year, Delta reported profits of more than one billion dollars, and nearly doubled that in 2013; the surcharge did not appear to be hurting business.

And there it was on that list, the flip side of Delta's success: The largest US–Europe carrier and my own preferred airline (for no other reason than that they or their partners fly to where I need to go) had declared it was the emitter of 4,668,157 metric tons of greenhouse gases. I'd accumulated a lot of miles on Delta—and thus a carbon footprint of more than a ton, I'm sure—over the course of reporting this book. What was I to do? Not fly? Not plausible. But pay an extra charge to more accurately reflect the cost of all that flying time? It seemed like a pretty slight addition given the stakes.

The charge will probably rise, slowly, over time. It was a reasonable concession to the single undisputed fact of my flight, that keeping my plane aloft came with a payload of greenhouse gases. It was not a moral question, it was pure economics: I would pay full freight, I would not shuttle the collateral costs of my flight onto you or anyone else.

Meanwhile, far from the diplomatic maneuvering and rhetoric of the courthouse, the question of how to keep airplanes aloft while minimizing their greenhouse gases was becoming a priority for some of Europe's leading air carriers. British Airways announced it would reach a goal of cutting its greenhouse gases 50 percent from 2005 levels by 2050 and opened a new research center ten miles outside London to develop new fuels extracted from biowaste. Virgin Atlantic's pioneering CEO, Richard Branson, concurred, launching an effort to devise fuel from recycled industrial waste, which he predicted would be usable for jets by 2014. In the summer of 2012, KLM flew a plane from Amsterdam to Paris powered exclusively with biofuels extracted from sugarcane and beet reeds. Lufthansa announced it would begin intensive research on developing fuel from algae and municipal solid waste in time for the opening of the new Brandenburg Airport in Berlin at the end of 2013. Other carriers, like Emirates and Singapore Airlines, announced that they would meet the guidelines established by the Europeans to at least bring their emissions down to 2005 levels by 2020. Even a few US carriers got into the act: While fighting the European requirement, United Continental announced it had launched an effort to devise a new family of algal-based biofuels because it would subject them to less dramatic price fluctuations than petroleum.

There was, meanwhile, another major American player, the largest fuel consumer in the country, which was aggressively pursuing research and development of less greenhouse-gas-intensive fuel—the US military. The Air Force Research Laboratory has been working for years on developing a reliable supply chain for fuel sources closer to home that are not subject to the expensive fluctuations of the global oil markets. They were certainly not motivated by concerns about complying with a European law, or concerns about climate change. The Department of Defense's interest is economic and practical: Each 10 percent increase in the price of oil, for example, means another $1.7 billion in fuel costs for the DoD.[15] They want a more predictable source of fuel closer to home. In the summer of 2012, as global tensions grew over the EU's aviation scheme, the Air Force sponsored a test flight of an A-10C fighter plane with fuel made almost entirely from alcohols extracted from the cellulose in wood, paper, and grass, and has now certified its use for parts of its fleet.

"The military would like to get away from being forced to rely on the highly volatile prices of the oil markets, and spur renewable energy technology in this country to make it more cost-competitive," commented Phyllis Cuttino, director of the Pew Project on National Security, Energy & Climate, which has been cooperating with the military in promoting the development of new fuel sources. Though climate change may not be a priority motivation for the Defense Department, it may be one of the subsidiary benefits of the department's search for more reliable alternative fuels.

So, while US air carriers were fighting the effort to impose stricter regulations on their use of greenhouse-gas-emitting fuels, the country's defense services were hungry for options to reduce reliance on foreign oil and, as what amounts to a bonus, reduce greenhouse gas emissions. The military effort shows one of the numerous co-benefits of switching away from reliance on fossil-based fuels, including a closer-to-home supply, price stability, less reliance on the Middle East, and reduced greenhouse gases. According to an assessment by the Air Transport Action Group, aviation biofuels have the potential to emit 80 percent less CO_2 per mile than traditional fossil fuels.[16]

(This figure, though, depends significantly on a thorough life-cycle assessment of biofuel source materials and on the process of fuel extraction. For example, fuels derived from corn, sugarcane, or vegetable oils—the three primary sources in the United States, Brazil, and the EU, respectively—involve significant alterations to land-use patterns, from growing plants for food to growing plants for fuel, and involve greenhouse-gas-intensive processing and transport. Other fuels, like those based on waste or algae, both of which are being researched intensively, suggest the possibility for biofuels to come with a far lower greenhouse gas load than either traditional fossil fuels or other biofuel varieties.)

But even the US military has been having trouble getting support for reducing its reliance on the unpredictable petroleum markets. The Pentagon has come under fierce Republican opposition in Congress when requesting authorization to purchase more expensive alternative fuels now in order to research and advance the technology to make them more cost competitive with oil in the future. There is a huge discrepancy, in fact, between the three to four billion dollars in annual subsidies and tax breaks provided to American oil companies (more on that in chapter 5) and the relatively small amount of support provided to the country's nascent biofuel research efforts, including those by the military. "In Washington, they've been having a fake 'political debate,'" said Cuttino. "Subsidies to alternative energy? You want to debate that? What about all the subsidies already going to traditional energy?"

Of course, when we talk about subsidies, we're talking about money from you and me, the taxpayers. Those "subsidies" include the generosity shown the commercial airlines in not requiring them to assess the actual costs of fuel for keeping their planes aloft. So the question really is: Do we subsidize the researchers seeking a way to develop less environmentally destructive fuels, or do we subsidize the ability of the airlines to continue utilizing the destructive fuels of the status quo? Either way, we'll pay.

———————

By the end of November 2012, the first test of putting a price on a flying carbon footprint was resolved: The EU succumbed to the global diplomatic and economic pressure campaign. It agreed to postpone implementation of the aviation law—a delay it dubbed Stop the Clock—in return for the Obama administration's commitment to work toward a greenhouse gas deal at the International Civil Aviation Organization.[17]

Tim Johnson, director of the London-based Aviation Environment Federation, a strong supporter of the EU's effort, characterized the retreat succinctly: "You had threats from every direction. The Chinese and the Indians were threatening non-compliance; and then you had the US threatening to not cooperate, that's something else entirely. That's what tipped the balance."

There was another factor, too: Shortly after the European retreat, I obtained a copy of a letter sent by the CEO of Airbus, Fabrice Brégier, to the chief of the Civil Aviation Administration of China (CAAC), Li Jiaxian. Airbus had been quietly lobbying the EU to rescind the measure after the CAAC had blocked the sale of forty-five Airbus planes to four different Chinese airlines. The deal was worth tens of billions of dollars. In the letter, Brégier reminded Minister Li about the postponement, and then pleaded for reactivation of the deal. They were in this together, he wrote: "Following this good news that EU ETS will not be applied to flights to and from destinations outside the EU, I hope that the CAAC will respond swiftly by granting approvals to the airlines for the purchase of Airbus aircrafts, so that these transactions can go forward without further delay. . . . Through our joint efforts," he continued, "we have managed to ensure that Chinese airlines are not unfairly impacted by the scheme as previously planned. I hope we at Airbus have been able to clearly demonstrate our strong support to Chinese aviation." Soon after, the Chinese purchase of the Airbus planes was completed.

ICAO quickly became the place for the refighting of old arguments. The United States returned to its position that the only emissions paid for should be those that occur within European airspace. In October 2013, ICAO kicked the can—committing, in principle, to the establishment of a global market governing greenhouse gas emissions from airplanes by 2020. It insisted that, in the meantime, no country could take unilateral action. The European Parliament responded by insisting that ICAO accelerate the market scheme to 2016, or it would revert to the original plan to subject all airlines landing in or taking off from Europe to emission regulations. The standoff continues.

So as of early 2014, Europe applies its greenhouse gas law to all carriers flying in and out of Europe, but has agreed to limit its authority to planes within European airspace. This "airspace approach" covers just 22 percent of the emissions generated over the course of international European flights, said Bill Hemmings of the Brussels NGO Transport & Environment. Which further bifurcates the global greenhouse gas strategy: You pay for emissions from your plane in Europe, but the minute you pass through the clouds above a line demarcating Europe from international airspace—or Europe from Russian airspace or Turkish airspace for that matter—no one is counting.

There was an interesting side effect to this epic tug-of-war. Hundreds of the world's airlines actually returned the emission allowances they'd obtained in 2013 to cover their 2012 emissions. So we can see, once again, another tally of airlines' greenhouse gases for the year after all that data was submitted to the FAA for review (and which it never claimed demonstrated any particular discrimination against American carriers). Now the data was in the form of allowances that had been obtained but not used, due to the diplomatic standoff—by our favorite airlines. According to records of returned allowances compiled by the EU's Directorate-General for Climate Action, the three major US carriers—Delta, United Continental, and American—came in for the year 2012 at just about a thousand tons less each than they had in 2011 (a reduction that is not statistically significant, out of a total for all three of twelve million tons). Nor does it appear that the US carriers actually rescinded those "fuel surcharges" they slipped into the cost of tickets—which means money that we fliers are supposed to be paying on each ticket for greenhouse gas mitigation is likely going to enhance the bottom line of the airlines.

It was easy enough to find the carbon footprint for my flight from San Francisco to Novosibirsk. Just plug the two cities into ICAO's own Carbon Emissions Calculator and, voilà, ICAO gives you the number.

Every step after that, however, means a collision of interests over how to account for that number—Fully? Partially? Not at all? Who runs the show? Things get tense when you try putting some meaningful muscle behind the number. And there is also far more at play than the mere number. Six bucks each way is not a debilitating figure any way you look at it. The battle over aviation emissions is at the cutting edge of a much larger battle: Establishing a price to pollute compels companies that rely on fossil fuels to acknowledge that they have, till now, been receiving a secret subsidy for years—in the form of environmental costs that they do not have to count. Ultimately, this sleight of hand is what the struggle over determining a cost for carbon is all about—lifting the veil on one of the oldest accounting tricks in the book, hiding the true costs.

It's the same, in one form or another, in every industry—including, as we'll discover in the next chapter, in agriculture and with the food we eat.

CHAPTER 2

Eat, Drink, Pray

Food: The Front Line

C limate change unfolds up in the majestic, remote regions above the clouds where those airplanes fly. Down here on earth, though, it gets down to work in very immediate ways. That's especially the case for farmers, who conjure food from three essential ingredients—the sun, water, and soil. Each of those elements is being profoundly altered by the changing climate. What we who live in cities register as freakish and sometimes welcome weather—more sun, less rain—can for a farmer be a matter of survival. Person by person, acre by acre, crop by crop, farmers are the earth's first responders.

There's another thing, too, about farmers' unique role in the climate feedback loop: They're among the world's leading contributors of greenhouse gases to the atmosphere. The cultivation, storage, transport, and distribution of food in our industrial agricultural system has one of the largest carbon footprints of any single industry, just behind energy refiners and transportation.

Greenhouse gases are emitted at every stage of industrial-scale food production. First there's methane, released through the waste stream of cattle. Next is nitrous oxide, NO_2, which is twenty-three times more potent a greenhouse gas than CO_2, and is primarily a by-product of nitrogen-based fertilizers, the use of which has doubled over the past fifty years. A team of scientists at UC Berkeley documented significant spikes in emissions during the planting season, when fertilizers are usually applied.[1] And then there's tilling of the soil, which unearths naturally occurring CO_2 and sends it into the atmosphere instead of keeping it in place for another millennium or two. Altogether, farmers are in the unenviable, though not totally unavoidable, position of being at the front lines of climatic shifts that they are helping to trigger.

So in our great fossil fuel experiment, farmers have been both a contributing variable and a fixed control. You can register the earth's changes through the experience of farmers, for the crops they plant are fixed in position while conditions change around them.

The state of California, it turns out, is an ideal marker for what is happening on the rest of the planet. The conditions here are more ideal for food growing than practically anywhere else on earth. The state was blessed with its geography, in the center of the temperate latitudes, with weather buffered by coastal cycles of fog and moisture from the sea. The state has rich, fertile soils and abundant biological resources, making it home to the most productive agricultural system in the United States, a thirty-eight-billion-dollar powerhouse of food that feeds the nation and the world. Now all those ingredients in California's geographically privileged rise are under stress wrought by the tumult in the atmosphere.

Charles Kolstad, an environmental economist at UC Santa Barbara, told me that California agriculture is being hit with a trifecta of converging forces prompted by climate change: longer seasons of extreme heat, shorter cold seasons, and dwindling water supplies.

That trifecta brings with it enormous costs—to farmers, to consumers, to governments. Figuring out just what those costs are, though, and who will pay what portion of them is fraught with the realities of biology, the unpredictability of water, the vagaries of the insurance industry, the uneven hand of politics, and, ultimately, the amount that you or I can afford to spend on, say, a tomato, an almond, or a cherry. To better understand where those costs come from, and to get a sense of what's really going on in the fields that feed us, I set off to visit farmers at the front lines of our great climate experiment in the summer of 2012. It was the hottest and driest summer on record in California, though it would take just one summer for that record to be broken. The costs were mounting. In the Golden State, conditions were shifting ominously beneath farmers' feet.

From my home in the San Francisco Bay Area, a trip into California's Central Valley is like a voyage to another planet. Here, sprawling broad and flat through the interior of the state, with the Sierras on one side and Nevada on the other, is the breadbasket of the country. While the vast expanses of the Midwest are devoted mostly to just a handful of crops—corn, soybeans, and wheat—farmers in the Central Valley grow dozens of different varieties of fruits and vegetables on a mass scale. They're the country's largest producers of peaches, plums, apricots, pears, walnuts, almonds, lettuce, tomatoes, spinach, and numerous other foods not to mention more exotic crops like persimmons and pomegranates. Their operations exemplify modern industrialized agriculture—heavily dependent on machines for planting and harvesting crops, on chemical fertilizers and pesticides to protect and feed them, and imported water from the north to sustain them.

In the San Joaquin, the biggest of the several valleys within the Central Valley, stretching four hundred miles from Sacramento in the north to Bakersfield in the south, cherries are a two-hundred-million-dollar-a-year business. In April and May, the rolling flatlands here are normally a time to celebrate as the cherry trees explode with the bounty of shimmering white and pink blossoms. The succulent fruits they deliver, in colors from yellow to plush red, are sold by the basketful. "Biting into a fresh cherry—there's no other experience like that on earth in my opinion," said Jeff Colombini when we met at his ranch during that spring harvest season. Colombini cultivates two hundred acres of cherry trees on land that has been in his family for forty years. He served three years as president of the Cherry Advisory Board. Colombini knows his cherries.

But this year, like the year before that, trouble was brewing under the blossoms. Colombini wanted to show me what concerned him about the season's crop and took me into one of his orchards. "These trees are like my children," he said as we walked between two rows of cherry trees just off the highway near Lodi.

To an untrained eye—mine—the trees appeared to be lit up with white blossoms. "Look closer," he said. Colombini pointed to the lower third of the branches; they had no blossoms at all. Then he pointed to the upper branches, which had blossoms just partially emerged from their buds. Only the branches sprouting from the middle of the tree showed the resplendent blossoms we normally associate with cherry season. His trees, he said, were showing "the stresses that come with not enough chill hours."

Cherries are one of the most highly sensitive tree crops; they come with a specific set of requirements that makes them the high-maintenance engines of specialty food crops. One of the key requirements is that the cherry trees get cold enough to go into a kind of hibernation in the winter months, preserving their metabolic energies for the springtime ripening that turns them into jewels for a spectacularly colorful but brief blossom and harvest season.

For a perfect California cherry, the trees need from twelve to fourteen hundred hours a year of chill time. Studies by agronomists at the University of California–Davis indicate that cherry growers in the San Joaquin Valley have been seeing an increasing number of seasons with from one thousand to eleven hundred hours of chill. "We're seeing those low chill effects every year now as opposed to how we used to see them, once every ten years or so," said Joe Grant, a UC Davis agriculture extension agent who has been advising tree farmers in the area for more than a decade. Scientists predict as much as a 40 percent decline in chill hours in the San Joaquin over the century between 1950 and 2050.[2]

Lack of chill leads cherries to ripen at erratic times—thus the half-opened buds at the tops of Colombini's trees—or, in some instances not at all, thus the nonexistent buds in the bottom third of his trees. And there may be another factor contributing to what's emerging as a crisis in cherry country: The fog is not coming like it used to, and it's the fog accompanying the cool air of fall and winter that helps shield young cherry buds from intense sunshine.[3]

The result, explained Grant: Cherries from this area of the state are shrinking in size, and the extended ripening time means that they are not as firm or deep in color when they are finally harvested. The yields are also dropping. And that, according to Grant, means that California is losing its competitive advantage. It used to be that California's warmer weather gave the state's cherry farmers a jump on their competitors in Oregon and Washington, where the cooler climate remains ideal for a bloom later in the season. Now in California cherry country, it's *too* warm.[4]

David Lobell, at Stanford's Center on Food Security and the Environment, co-authored a report in 2011 predicting just the sorts of dramatic declines in cherry yields and quality that have been seen across California as a result of the shifting conditions.[5] "The lack of chilly nights in winter," he said, "makes cherries one of the most vulnerable to climate change." Packinghouses in the area reported a dramatic fall-off in orders as yields declined in area farms in 2011 and 2012. In 2011, California cherry growers received a record $22.5 million in crop insurance payouts—sending crop insurers into the red. For each $1 paid into the program, $1.60 was paid out. That gap was covered by the USDA—us taxpayers—which paid some eight million dollars to subsidize the losses. The 2013 season was about as bad, with half the yield that had been considered normal before 2010.

Colombini told me he's not sure how long he'll be able to keep growing cherries in the way he does now, and his father did before him. "I don't know what the causes of the climate change are," he said. "But I would definitely say that our climate has changed.

The bottom line is, we don't even know what a normal season is anymore . . . It's crazy, but our weather is becoming very much more variable than it had been in the past. That's what's going on."

Cherries are the canaries in the climate mines. Those half-grown blossoms send a signal of more trouble to come.

In late 2011, California governor Jerry Brown shared his fears of climate change with a group of policy makers, businesspeople, and scientists. From the stage of San Francisco's Academy of Natural Sciences, the governor cited a philosopher rarely, if ever, quoted by American politicians. "We want to avoid a Hobbesian situation," he declared, "the brutishness that happens as things get tighter." The audience chuckled uneasily. Thomas Hobbes, who saw society as a clash of individuals driven by immutable self-interest, and government as an instrument of restraint upon those selfish instincts, is not a philosopher who leaves you with a warm, optimistic feeling for the future. What Brown feared the most, he said, were the multiple billions of dollars that would be drained from the seriously strained state budget to deal with the accelerating consequences of climate change and the clashes that would create among the state's myriad powerful interests—which he and other political figures would be compelled to mediate.

Hobbes, a seventeenth-century man, was at his prime about one hundred years before the industrial revolution, which is when scientists say that greenhouse gases were first emitted in quantities large enough to change the fundamental atmospheric balance. They've been accumulating at a faster and faster pace ever since. Hobbes could not have imagined the changes wrought by the steam engine and beyond. But the consequences of those changing conditions, as Brown pointed out, are straight out of Hobbes: Atmospheric chaos looms, and as it does so, conflicts arise over increasingly stressed resources. Nowhere is this dynamic more clear than in the panorama of pressures that climate change creates for farmers—dealing as they do with the weather and its variables every single day. Discovering Hobbes at play in the fields of California was not difficult.

About sixty miles south of Colombini's orchard, I took the exit off Highway 99 for the tiny town of Hughson, and headed to a place that has the ear of farmers and is well positioned to respond to the changing conditions they face—the Duarte Nursery, the biggest commercial nursery west of the Mississippi River. Located just south of Modesto, the nursery was founded in 1974 by John Duarte Sr. to develop wine grapes for the southern part of the state. Since then it has thrived, growing seeds for thousands of farmers here and throughout the Pacific Northwest. There's a good chance the fruits, berries, and nuts eaten west of the Mississippi begin their journey to market as seedlings at Duarte's thirty acres of greenhouses, labs, and growing plots. Their specialty is using state-of-the-art breeding techniques to tailor varieties to the specific conditions faced by farmers. John Duarte Jr., who took over from his father as president of the nursery, told me that those conditions get down to three basic elements: heat, water, and salt. The nursery, he said, has been trying to keep pace with the dramatic changes they've been seeing in all three.

Duarte, like Colombini and many other farmers in the state, senses the changes but is hesitant to identify the cause. "Whether it's carbon built up in the atmosphere or just friggin' bad luck," he told me, "the conditions are straining us."

A broad spectrum of scientists and policy makers are convinced that the changes under way in the Central Valley, as well as other food-growing regions of the country, are linked to the disruptions wrought by greenhouse gases. The California EPA issued a report in August 2013 suggesting that climate change posed huge risks to the state's continued growth. Later that year, the state's Department of Food and Agriculture convened a panel of experts to propose ways that farmers of specialty crops, like cherries, should begin to adapt—now—to climate change. Their recommendations included greater protection for pollinating honeybees, enriching the soil with natural composts to create greater resiliency and water absorption, and a host of other measures that will challenge farmers to take a different—and ultimately more sustainable—approach to their everyday operations.[6] By January 2014, the state's drought, which prompted Governor Brown to declare a statewide emergency, had sparked the attention of even President Obama, who appeared in Fresno, the largest city in the Central Valley, to announce a $183 million aid fund designed to help farmers devastated by the dwindling supply of water.

Like everything in climate change, the consequences unfold like a sequence of falling dominoes.

First there's the temperature, a jagged progression over the last decade of unusual highs trending upward, as well as unexpected lows occurring at times of the year that debilitate growing crops. The federal EPA estimates that valley temperatures will rise between 1 and 3.6 degrees Fahrenheit by 2050.[7] One result: Less snow is falling in the Sierras, and less snow means less water melting for farmers in the dry months when they need it most. And then there's the salt, which is building up in the fields because of the lack of fresh water to drain it out of the soil as well as the steadily rising level of the Pacific Ocean. One set of phenomena cascades into the other. Farmers may not agree on the causes, but they have no choice but to confront how those cascading forces are changing the conditions for growing food.

Duarte is tall, stocky, genial. He took me to the nursery's laboratory inside a low-slung brick repository for the company's seedlings. Through Plexiglas, we looked at what felt like a science-fiction glimpse into the future of food adapted to climate change. Here an experiment that Duarte could control was under way. Four women were sitting in a bacteria-free compartment, with their hair bundled in antiseptic caps and their hands fitted into surgical gloves. They used tweezers to insert pinkie-sized sprigs of apricot, peach, and almond trees into tiny pots of enriched soil. Within weeks, the high-speed growth sustained through a specially designed agar mix of nutrients ensured that Duarte would know which rootstocks respond best to the higher temperatures, diminishing water supply, and increasing salt content that are rapidly becoming the new growing conditions for Central Valley agriculture.

A study sponsored by the state Energy Commission, co-authored by Kolstad, the UC economist, predicted that over the next forty years, assuming a two-degree Fahrenheit temperature rise over the next four decades (a conservative assumption), yields of California citrus crops would drop 18 percent, grapes 6 percent, and cherries and other tree orchard crops 9 percent.[8] The study focused on tree-based perennial crops—Duarte's specialty—because they are fixed in place over a generally twenty-five- to thirty-year life span, and cannot easily be adapted to changing conditions. Trees can't migrate or be moved to more conducive climates. A 2013 assessment by the USDA suggested that crops across California would be subject to intensifying stress due to hotter temperatures and declining water resources, with ripple effects across the spectrum: increasing vulnerability to a new array of heat-adapted pests, alterations to pollination patterns from increasing heat, impact of diminishing low-chill factors for cherries, blueberries, apples, and other annual specialty crops, which it predicted would lead to declining yields.[9]

Duarte is working as quickly as he can; the changes are happening more rapidly and dramatically than previous climatic shifts. "There's a limit," Duarte conceded, "to what we can do with genetics."

Millions of dollars are being spent by Duarte's team of botanists, as well as a network of other nurseries and colleges in the area that he called "the Silicon Valley of agricultural innovation," to help California farmers adapt to the rapidly changing conditions for their crops. Some of them, unlike Duarte, are working on new genetically modified varieties—though the technology is not yet in widespread use on crops in the state (as distinguished from the substantial cultivation of GMO corn and soybeans in the Midwest). All are scrambling to slow the corrosive effects of the changing climate on crops.

The most acute of those effects can be found in the central ingredient of their trade, delivered via aqueduct from the north to the south—water.

Irrigation to the Central Valley is not expected to come anywhere close to what farmers say they need over the coming decades, according to the state's own report on the future reliability of California's water supply.[10] Over almost a century, the California Department of Water Resources (DWR) found there has been a 10 percent decline in runoff from the Sierra Mountains during the critical months from April to August. That runoff feeds into the San Joaquin and Sacramento Rivers and ultimately into the Sacramento Delta and the seven-hundred-mile network of pipelines and aqueducts that carries water south to nourish crops and supply fresh water to drier parts of the state. There are actually two systems in California's labyrinthine maze for delivering water: The state water project generally supplies cities and municipalities; and the federal water project supplies farms. Both draw from the same supply—water that falls in the north and follows the course of rivers until ending up in the delta, where it is shunted by massive pumps southward. But over just the past decade, the water runoff has declined at double the rate; now it's 20 percent less than it was in 1910, according to John Leahigh, chief of operations planning for the water department. For the three years between 2006 and 2009, Leahigh said, the runoff amounted to the equivalent of two "normal" years. It appears the state has a new normal. In 2013 and into 2014, the state experienced a major drought—more severe than any since records have been kept.

"We know California is getting hotter and drier," said Scott Loarie, at the Carnegie Institution for Science at Stanford University. "Getting water to your crops is getting to be a bigger deal. While the temperature effects are significant, the biggest issue is really climate seen through the lens of water. That's where we have the greatest vulnerability."

It's not only the volume of water that's changed over the past century; what's also changing is the form it takes while falling from the sky. Rising atmospheric temperatures mean that water in winter no longer comes as snow, but as rain. As skiers have discovered, the high Sierra snowpack is shrinking, and along with it the state's natural reservoir. The snow has long served to preserve water in frozen form until the spring and summer melts when rainfall drops significantly in the south. But snow isn't falling like it used to: The state EPA predicts a 25 to 40 percent decrease in snowpack water storage capacity by 2050.[11]

As a result, millions of acre-feet of water are overwhelming the system designed to catch and store it. (Large volumes of water are commonly measured in acre-feet, which means the amount of water it would take to cover one acre one foot deep.) The reservoirs and canals were built according to precipitation patterns of the 1950s and earlier, when the snowpacks melted just in time to provide the southern part of the state with water. When rain falls instead of snow, massive quantities of water are wasted, if flowing into the ocean can be described as wasted, because they're flowing at the wrong time. "There's less water coming into the system during the spring and summer when the farmers really need it," said Francis Chung, chief of water modeling for the DWR.

Many valley farmers tend to blame their water troubles on a court order requiring the state to reserve a portion of the fresh water of the Sacramento Delta for the preservation of threatened river fish. But that accounts for no more than 15 percent of the delta water supply, and is drawn from a pool, said Chung, "of less water for everyone." Eighty percent of California's water is used on farms— the remainder is fought over by cities and all other non-agricultural users—so when there's less water for everyone, farmers feel it.

Rising sea levels exacerbate the problem. According to the National Academies of Science, the sea level in the San Francisco

Bay is expected to rise from seven to eighteen inches by 2050, and close to triple that by 2100—threatening to inundate with salt the vast, mulchy, richly biodiverse Sacramento Delta. To keep the salt water out, the state now diverts massive amounts of that fresh water from the mountains to act as a kind of liquid barricade. About two hundred thousand acre-feet of fresh water are needed to hold back each foot of the rising ocean—and every one of those acre-feet would otherwise be used to satisfy the thirst of California's farms and cities. There's even a point identified by the state where the fresh water from the mountains confronts the salty waters from the sea in the middle of the Sacramento River. It's called the x-line.

I went out searching one summer afternoon for the x-line along the roadways of the delta. According to state maps, there it was, just off the riverbank from the town of Antioch. The murky water meanders as slowly here as it does elsewhere on the river. But just underneath the surface is where California has established its Maginot Line against the intrusion of salt into the delta, where it fights most directly and aggressively, day in and day out, the consequences of climate change for the lifeblood of California agriculture.

The fresh water used to restrain the sea's intrusion and enable the delta to meet state salinity standards, according to Francis Chung, would otherwise be diverted into Central Valley irrigation canals. "It takes more water [every year] to maintain the current salinity standards," said Chung. The more the sea rises, the less water is available for farmers growing food. "Instead of giving you, say, one hundred acre-feet [of water], we'll be able to give you ninety. What do you do with that ten? You use it to repel that intruding sea salt." In other words, he said, "We are trading quality of the water for quantity."

And there's yet another twist: Despite the gargantuan fresh-versus-seawater struggle in the Sacramento River, the Maginot Line is not exactly made of concrete; in fact, being entirely made of water, it's quite porous. Some of that seawater escapes through the freshwater barrier. Forty railroad cars' worth of salt—about five hundred thousand tons a year—already flows daily out of the delta into the fields of the Central Valley, adding to soil already made salty by pumping groundwater from what millions of years ago was the ocean floor.[12] According to the Water Institute at the University of California–Davis, the salt content near the federal and state pumps that send water to the Central Valley is expected to rise by 4 to 26 percent over the next four decades[13]—a number dependent on many variables, including how much the sea actually rises and how much fresh water is available to restrain the ocean's ebb into the delta. "It's getting harder and harder to keep the salt levels at the right level," commented Jeffrey Mount, a UC Davis hydrologist and member of the Delta Independent Science Board.

Some farmers in the western valley are being forced to adapt by switching from salt-sensitive crops like strawberries and avocados to less sensitive—and less profitable—crops like alfalfa and wheat, according to Daniel Cozad, president of the Central Valley Salinity Coalition, a group of local farmers, businesspeople, and government officials. "Unfortunately," he said, "the higher the value of the crop, the more sensitive it is to salt." A study by UC Davis estimated that if salinity continues to rise at the current rate, by 2030 the financial costs to the Central Valley could be huge: As much as $1 to $1.5 billion a year in decreased agricultural activity, amounting to some twenty-seven to fifty-three thousand jobs.[14]

The Errotabere family's farm sits about twenty miles north of Duarte's nursery. They've been growing food here since 1948 when the family moved from Spanish Basque country and purchased 860 acres near Turlock, a farm town located practically in the center of the Central Valley. Now it spreads over five thousand acres, overseen by two Errotabere brothers and a sister, who run what might be politely called a family-run factory farm: rows upon rows of tomatoes, garlic, onions, lettuce, melons, and the most valuable of all—640 acres of almonds.

John Errotabere, modest and precise, manages the finances for the farm. He greets me in a flannel shirt and well-pressed khakis, and ushers me into a grove of those prized almonds. Acre per acre, they are far more profitable than any other single crop on his farm. They also devour water—almonds require double the amount needed by tomatoes, for example, and need it year-round.

In February every year, the Bureau of Reclamation, which runs the federal water project—the primary source pool for agriculture—does an assessment of the projected rainfall, the depth of the snowpack, and reservoir levels to determine how much water is available to be channeled to Central Valley farms. Farmers have grown accustomed to erratic water flows—over the past decade, they've hovered at between 50 and 70 percent of baseline levels established after creation of the project in 1933. But just two months after Governor Brown invoked Thomas Hobbes, the bureau reduced its yearly allotment for valley farmers from 50 to 40 percent, an unprecedented reduction that late in the growing season.

When Errotabere heard the news, he responded by looking not up toward the Sierras and their lost liquid bounty, but down. He installed pumps to draw water from 350 feet below, just as his grandfather had done decades earlier. Here above what was once the ocean floor lies an aquifer that had provided water for generations of farmers before him.

But this time, the news came dribbling up from the wells with a message from the center of the earth: His family's ranch had nowhere near the water resources he thought it had. The trickles of water were tainted with high levels of boron and salt. It was unusable, he said, for his crops. The aquifer was drained by, as Errotabere put it, "too many straws sucking on one drink." It rapidly became clear that there would not be enough irrigated water to sustain both those highly profitable almonds and other crops on his farm. "With those almonds, you know, it's not like we can change our minds. That investment takes years to recover." Nor, being trees, can they be easily moved or substituted.

Errotabere was faced with a brutal decision: Which crop to sacrifice in order to sustain water for those highly profitable almonds? He remembered the month: In February 2012, he'd just finished planting 160 acres of tomatoes with an already existing contract for delivery to a nearby processing facility. Errotabere decided to run a tractor over his tomato vines, churning them back into the soil. Then he decided not to go through with planting onions and garlic in a 340-acre field that he had just fertilized but not yet seeded.

Overall, Errotabere had to take 10 percent of his five thousand acres out of production in 2012. That cost him, he estimated, about $450 an acre—more than $700,000. He expects to take far more acres out of cultivation over the coming years if the water shortage continues—and most state estimates anticipate it will.

Trade-offs like Errotabere's are destined to become more dire as climate change exacerbates the stresses. In 2013, during the state's record-breaking drought, the situation grew even more intense: The allotment plunged to a disastrous zero, compelling farmers to fallow hundreds of thousands of acres and Governor Brown to declare a water emergency.

Hobbes would have a ball with California's water system—which requires ever more intervention by the government into the complex thicket of interests with a claim on the state's water. Decisions have to be made: How much fresh water should be preserved to hold off the advancing Pacific? How much to maintain the delicate ecological balance, including the fate of several endangered marine species, in the Sacramento Delta? How much should be reserved for farming, how much for city dwellers? When in 2012 the irrigation district in Modesto agreed to sell twenty-five million acre-feet of its water allocation to the city of San Francisco, which is facing concerns over its water supply from the Hetch Hetchy reservoir, there was outrage across the valley. And when Modesto finally pulled out of the deal in 2013, due to opposition from farmers who claimed the deal would leave them dry in drought years, San Francisco was left to find a replacement source, at a cost far higher than the seven hundred dollars per acre-foot it was willing to pay the Modesto district—if the water is available at all.

Or what happens when cities in one state are pitted against other cities in another? This is the calculus now at play as farmers in California's Imperial Valley, aligned with the city of San Diego, assert that Arizona has been diverting too much of their water from the Colorado River.

Or what about the vintners of Napa Valley, who are contending with decisions between planting their world-famous varietals of Cabernet Sauvignon, which like cool weather, or more heat-tolerant varieties of Chardonnay and Pinot Noir? Over the next thirty years, the temperatures in Napa are expected to rise by at least 1.8 degrees Fahrenheit—an increase of great import to wine growers who plant their vines for twenty-five-year spans of grape production. "We don't want to lock in mistakes," said Stuart Weiss, a conservation biologist who consults with wineries on how to adapt to the rising heat and diminishing water in the country's premier wine-growing region.

Or what trade-offs are made when crop is pitted against crop—which is how the tensions over water often play themselves out, as they did with Errotabere, on each farm, farmer by farmer, exercising their own triage?

One after another, Central Valley farmers are making a similar calculation—to sacrifice one crop in favor of another, or in favor of none at all. Across the Central Valley, more than one hundred thousand acres were fallowed in 2012 due to a lack of water. By 2013, more than half a million acres were taken out of production. The costs to farmers, laborers, and other parts of the agricultural service economy—seed companies, processing facilities, truckers—from land taken out of cultivation due to lack of water quickly jump into the tens of millions of dollars. Add to that the multiplier effect: Every dollar put into agriculture generates five dollars to the local economy. Official unemployment in the valley hovers around 15 to 20 percent, the highest in the state. During the 2013 drought, it jumped in parts of the valley inhabited mostly by farmworkers to as high as 40 percent. Studies by the Economic Development Administration show an inverse corollary between water supplies and unemployment: The lower the former, the higher the latter.

(In the spring of 2014 came a twist to the parching of the Central Valley: on its western edge, the part most acutely impacted by the drought, the local water district proposed the development of a "solar park" on thousands of acres of fallowed farmland—suggesting that if the water supplies continue to decline, the sun may offer some economic hope.)

Meanwhile, the price of water keeps rising. Near Los Banos, a town in the northern part of the San Joaquin Valley, Berj Manoukian has been totally reliant on irrigated delta water since a couple of wells on his property came up dry or loaded with too much salt and boron to use. He doesn't grow trees; he grows melons on a thousand acres, as did his father and grandfather. I visited his farm during harvest season. Row after row of vines stretched into the distance and, every foot or so, there was another plump cantaloupe. He cut one off the vine, sliced it quickly into perfect orange-yellow quarters, and handed me one. "Terrible!" he joked. They were delicious, juicy and ripe. This is the highest moment of satisfaction all year, he said; the rest is pretty much nonstop pressure.

When the word came down from the state capitol about a 20 percent reduction in his federal water allotment, Manoukian knew he'd have to purchase private water in its place. The price of that water has more than tripled over the past decade—from $100 to $350 per acre-foot, he told me. Each 10 percent drop in his water allocation means another ten- to twenty-dollar-per-acre-foot rise in his price for water—an immediate jump of at least thirty thousand dollars for the whole farm. Instead, Manoukian opted to fallow a third of his land over the past five years, because the economics don't make sense any longer. That means hundreds of thousands of dollars in lost income to Manoukian, and lost cantaloupes for us.

I asked Manoukian if he'd seen the film *Chinatown*, that epic tale about the 1930s battle over water between the farmers of the Owens

Valley—located some hundred miles to the southeast—and the cabal of Los Angeles real estate developers. His response surprised me: "No, haven't seen it," he said. "Don't plan to. People around here think we're the next generation of Owens Valley farmers."

In the film (and in reality, for the film was based on the real back-room deals of the 1930s), the Owens Valley farmers lost their water rights to unsavory developers. Today Manoukian sees himself and his fellow farmers as the lone wolves on the edge of another sleight of hand over water. This time, however, it's not the diversion of water from farmers to housing moguls, as in *Chinatown*, but the shrinking supply all around. As Governor Brown ruefully suggested, this creates an ever more tense struggle among those who can claim legitimate entitlement to the state's water supply. Cities, industries, environmentalists, and farmers—in other words, some of the state's most powerful interests—are fighting over a dwindling resource. "No one," said UC Davis hydrologist Jeffrey Mount, "should be surprised that water equals money, whether it's growing housing or growing crops."

That equation has certainly not been lost on Barat Bisabri, who farms almonds, pomegranates, and grapefruit near the town of Westley. His farm is located about an hour north of Manoukian's, and you can see his fields of almond trees as you speed by on Highway 5. Bisabri has a PhD in entomology and spent twenty years working for Dow Chemical, helping the company launch a line of insecticides designed, he said, "to be far more efficient and less toxic to anything other than the bug we wanted." After retiring from the company, he and a group of friends bought a thousand acres of this prime farmland. He had been hoping that aquifer would provide backup to his unreliable, irrigated water supply.

In 2009, facing an allocation shortfall from the federal water project, Bisabri was compelled to make a decision: take his chances that the water below his property was pure enough for his 160-acre almond orchard, or just let the trees die. He opted for the water below. But that was tainted with salt.

Three years later, he took me on a walk through his poisoned orchard. Bisabri had, unintentionally and unwillingly, been testing the impacts of salt on almond trees, another casualty of the water squeeze. He pointed to the brown, shriveling leaves in his once robustly healthy grove. The trees were dying, their growth and maturation stunted by the salt that came up with the water, drastically reducing their ability to absorb nutrients. He pulled off one of the few almonds that were produced by the dying trees and held it in the right palm of his hand; in the other hand he held an almond from a nearby orchard watered from different and far less salty sources. The almond on the right was about half the size and a pale brown compared with the one on the left, a fine oval in the dark brown color of healthy almonds we're accustomed to eating.

In October 2012, Bisabri bulldozed that entire field of dying almond trees, which cost him more than a hundred thousand dollars. He was going to plant a fresh orchard, and had purchased new rootstalks promoted as salt-resistant.

This is how the domino effect works: Bisabri's almond field, like those of other farmers who grow food far away from the mountain peaks where the water falls from the sky at the wrong time, was the final falling domino.

Bisabri confessed that he was nervous about the uncertainties ahead: "I'm just hoping that the predictions are wrong, the predictions of climate change and its effects in the Central Valley. Otherwise, what am I going to do?" His question just hung there in the hot, dry air.

Alas, few of the predictions about the impacts of climate change in California food country have yet been proven wrong. The state's EPA predicts rising temperatures, increasing salt, and further stresses on the delta's ability to supply the Central Valley with adequate amounts of water without massive investments in new dams and reservoirs.

As I made my forays into the Central Valley, a drought was break-ing out across the United States. July was the hottest month across the country since records were kept. Water supplies in the Midwest collapsed; at least half of the corn crop in Iowa, Illinois, and Indiana was wiped out. Cows are the primary consumers of midwestern corn, which means that all those cattle farmers across the country, from California to Nebraska to Texas and North Carolina, were suddenly paying 10 to 15 percent more for cattle feed, according to the USDA. The situation had reached the point that by early 2014, President Obama launched seven "climate hubs" in highly stressed agricultural regions across the country: From Oregon to Iowa to Oklahoma and New Mexico, the new hubs would help farmers adapt to and respond to the droughts, new pests (which have been following the warmer conditions), floods, and fires that are being experienced as climate change accelerates around the country.

The disruption of normal patterns in California and across the United States is reflected in similar climatic zones around the world. In the Mediterranean countries, yields of corn and cereals—includ-ing the wheat for Italian pasta—have been stagnating or declining over the past decade, according to Jorgen Olesen, a professor of Agroecology at Aarhus University in Denmark, who co-authored a comprehensive survey of how climate change is altering European agriculture.[15] He found that while some regions may benefit from warming—for example, vegetable farms have opened in Denmark, which would have been inconceivable five years ago—those benefits come nowhere close to outweighing the negative impacts on yields in the agricultural heartland in the center of Europe. "What we're seeing," he said, "is cold winters followed by very hot, dry summers—extremes on both ends."

Go pretty much anywhere in the world, he added, and climate change is putting new stress on food production. In China's Northern Plains, the center of that country's food production, groundwater levels are declining by as much as three feet per year and are not being replenished, due to ever-sparser rainfall, leading to severe drops in yields for wheat and maize. Half the world's people live in countries where water tables are falling, the Earth Policy Institute concluded in 2012. Most of those are in Africa, where disastrous droughts and desertification have led to plummeting food production. The Intergovernmental Panel on Climate Change (IPCC) warned in 2014 that a food crisis looms right behind a perfect storm of lower rainfall in already dry areas, and more torrential rainfall in areas that are already wet. Assuming current emission levels, they predicted that every decade through the rest of this century the global population will increase by as much as 14 percent while the rate of agricultural production decreases, decade by decade, by as much as 2 percent.

"The volatility we're experiencing in California is really a microcosm of what's going on globally, which is [that] extreme weather is going to be driving up the cost of producing food," commented David Freidberg, who runs one of the nation's biggest private crop insurance companies, The Climate Corporation. Climate change did not create food shortages; those can be as much a result of skewed distribution systems as particular changes in the growing conditions. But what it has done is exacerbated the huge gap between those— us—who are well fed, and those who are struggling to eat. Stresses on food production, said the IPCC, are already one of the most telltale impacts of climate change, and are likely to become a major contributing factor to global instability in the coming decades.[16]

Back in California, I visited the single most significant fixture in the state's complex water system. In a huge concrete structure near Tracy, on the edge of the Sacramento Delta, I stood at the edge of one of nine enormous pumps that send a million gallons of water a minute from the delta down south to the Central Valley. Churning like horizontal propellers, the pumps—called impellers— are managed by the federal Bureau of Reclamation. The water that arrives here from the Sierras up north is pumped over a quarter-mile, thirty-degree rise onto a hundred-mile journey south to the fields of the Central Valley. I asked Pete Lucero, communications director for the bureau, what the rising sea level and other pressures would mean for Central Valley farmers who are dependent on the water passing through these pumps. His response suggested just how brutal things will get over the coming years.

"I'll tell you exactly what it means," he said, over the steady hum of the massive pumps. "Someone's going to feel the effect if we have to move water that we don't technically have to someplace else that doesn't have it. It has to come from somewhere. There's only so much water in the ground. There's only so much water in the Sierras. And there's only so much water falling from the sky . . . You're going to have losers in every part of the state of California. Cities and towns that rely on this water will lose. Farmers in the Central Valley are going to lose. Farmers in the [Sacramento] delta are going to lose. Environmentalists are going to lose. That's what it means: Everybody is going to lose."

That was pretty sobering news. It would certainly get a deranged chuckle out of Thomas Hobbes. So if everybody loses, who pays?

CHAPTER 3

Earth, Wind, and Heat

Food: Our Liability

From the fields of California, we travel briefly to the eye of the financial storm triggered by climate change—the nation's capital. From behind the granite pillars of the Government Accountability Office (GAO) comes the beginning of an answer to the question of who pays for climate-change-induced damage to our agriculture. The GAO was established by the US Congress in 1921 after expenditures during World War I wreaked havoc on government finances. The new agency was given independence from the executive branch to provide ongoing assessments of the performance of federal programs. Inside this warren of hallways sit the seasoned accountants and auditors of the US government.

Since 1990, the GAO has been providing biannual reports on the greatest threats facing US government coffers. In February 2013 they issued their latest 275-page compendium of alerts that require federal attention. To a list of "fiscal exposure" risks that stretches from terrorism to food safety to aging transportation infrastructure, the GAO added, for the first time, climate change. Specifically, it said that the two areas of greatest financial vulnerability—both requiring a more "proactive" approach—are the looming costs to the federal system of crop insurance and flood insurance.[1] We've heard much about the national financial impacts of rebuilding from the floods in the wake of hurricanes like Katrina and Sandy. But we don't hear much about the costs borne by the nation when climate change wreaks havoc on the food system. Those costs come in the form of a government guarantee to private insurance companies that offer protection to the high-risk profession of farming—aka, crop insurance.

Farmers have always been gamblers, betting on the rain, on the temperature, or on holding on to a crop instead of selling it in the hope that the price will rise because of those factors somewhere else. Climate change shifts the odds against them. "There's always going to be good years and bad years in agriculture," commented David Lobell of Stanford University's Program on Food Security and the Environment. "Even in the future, there will be some good years. But those will become relatively less common, and what we now consider a bad year will become relatively more common." The uncertainties farmers face in committing their labor, technology, and inputs long before the revenues roll in at the end of the growing season are hedged with insurance, a system that was invented at a time when climatic extremes were understood to be grievous aberrations from the norm. Now it's becoming the ultimate backup for farmers as aberrations become the norm.

Crop insurance, one of the great, lasting legacies of Franklin Roosevelt, was born out of the first great environmental catastrophe to hit American farmers—the Dust Bowl of the 1930s. The raging winds that swept across Oklahoma and Texas wiped out the livelihoods of hundreds of thousands of farm families, left with no backup in the days before the New Deal. Moving was their only option. They became America's first environmental refugees. The mass displacement of Oklahoma farmers was later immortalized by John Steinbeck in his classic novel *The Grapes of Wrath*. Roosevelt's Secretary of Agriculture Henry Wallace, himself a farmer, created a program in 1938 in the hope that farmers would never again be wiped out by freakish turns of the weather.

Seventy-five years later, we have a new term for people fleeing dramatically altered growing conditions—climate refugees. We imagine such refugees as the residents of distant islands being inundated by the rising seas. But that phenomenon could be much closer to home than we think. A World Resources Institute map identifies areas of major environmental stress due to alarming drops in the water supply in agricultural regions across the earth. Blazes of red, signifying the highest danger, are splotched across the planet—in large portions of India and northern China, the coast of North Africa, much of southern Europe, and central Mexico. And there it is, a red splash across the American Midwest into Canada, and, finally, a spot in the center of the state of California—the Central Valley.[2] Now take that visualization of risk and place it atop another illustration, this one a color-coded map from the USDA's Risk Management Agency (RMA), of the country's crop insurance indemnities county by county in February 2014. The colors range from white (no indemnities, which generally means no farms) to light yellow to red to dark brown, in shades of escalating exposure. Of the thousands of American counties, two in the Central Valley share the distinction, along with a broad swath of the Midwest and the Great Plains, of being in the darkest brown, representing those with more than ten million dollars in insurance obligations.[3] The consequences of climate change identified in the first map are being shouldered, at least in the United States, by the federally subsidized crop insurance system articulated in the second map.

Crop insurance is all of our investment in farmers' ability to continue producing our food, and staying put. It's turning out to be their—and our—primary hedge against the risks of climate change. The system relies on a complex ladder of escalating commitments by the federal government, involving a tangle of private and public interests. While the insurance is administered through private companies, taxpayers subsidize farmers' premiums, the insurance companies' operating expenses, and, when losses occur, payouts.[4] For basic catastrophic coverage, the government pays 100 percent of the premiums (farmers pay three hundred dollars in administrative costs) which offer payouts to cover up to

55 percent of revenue losses off a baseline measured from previous performance. Farmers can obtain additional coverage specifically tailored to particular conditions—rain, hail, or drought, for example—or to hedge against declining prices by guaranteeing a certain revenue level up to 85 percent of losses. For those policies above the basic catastrophic coverage, the government pays an average of 60 percent of the premium costs (the higher the coverage, the more substantial the government contribution). The government also reimburses insurers for damage payouts ranging from 20 to 65 percent, depending on the type of policy, amount of damages, and location (the rates differ state by state). This is intended to ensure that the insurance companies sustain a minimum 14.5 percent yearly return on their investment[5] (though it's frequently been far more than that). It's almost impossible for the companies to go bust no matter how bad the year, because the government picks up increasingly significant portions of the tab. Thus, the government has a deep and expensive interest in the financial health of farmers, which is being threatened season by season by the increasing volatility of the weather.

Before 2010, the concept of climate change was barely acknowledged officially by the USDA. But little by little, as the risks and financial burden grew, the lexicon of climate change was nudged into the USDA's vocabulary. In that year, the department's Risk Management Agency, which oversees the crop insurance program, commissioned an internal report suggesting the huge climate-based risks ahead for farmers.[6] They found that California was one of the most vulnerable parts of the country. "California produces 95% of the United States' apricots, almonds, artichokes, figs, kiwis, raisin grapes, olives, cling peaches, dried plums, persimmons, pistachios, olives and walnuts. Since the production of these commodities is so concentrated into one geographical area the climatic impacts in these agricultural markets could be profound." Unreliable water supplies, the agency predicted, would continue to rise as a major stressor for farmers. Yields will decline, it concluded, and the property value of irrigated farmland—in other words, fields subject to water cutoffs—could drop "by as much as 40%."

By 2013, the USDA had expanded its research significantly and issued an official two-hundred-page report, in the same month as the GAO's "high risk" warning, that identified the broad spectrum of challenges presented nationwide by climate change to the future of American agriculture.[7] The agency identified areas of primary concern across a spectrum of changing conditions. While some areas of the country might benefit from warmer temperatures—for example, in New England and the Upper Midwest—the overall trend suggests more crop liabilities to come. "Continued changes by mid-century and beyond," the agency concluded, "are expected to have generally detrimental effects on most crops and livestock." Those impacts include increased costs for herbicides to control the spread of weeds accelerated through the combination of hotter temperatures and increased concentrations of CO_2; new pests attracted by warmer weather; and disease rates in cattle that rise due to higher humidity, creating happier climes for damaging bacteria. It also identified such factors as erosion: The increasing severity and frequency of storms could overwhelm the ability of the soil to absorb water, leading to increased erosion of precious topsoils in the runoff.

Lloyd's of London, the scion of the worldwide insurance industry, identifies similar trends as its exposure to climate-change-related events becomes a source of rising concern, as well.[8] "Lloyd's believes that the vast majority of natural perils are currently insurable," stated a company report on the risks of climate change. It then added, ominously, "Alternatively, if the pace of climate change grows faster than expected, this could change our view." Either way, the 150-year-old insurance company, which practically invented the large-scale hedging of risk, concluded that as a result of rising temperatures, "we foresee increased heat stress in livestock . . . [and] increase of damage to some crops."

As agricultural landscapes are altered, crop insurance rates, and payouts, are skyrocketing—from $4.3 billion in 2010 to $10.8 billion in

2011, to $17.3 billion total in 2012, the year of the Midwest drought.[9] That drought sent the corn and soybean crops reeling in Iowa, Illinois, Indiana, and elsewhere, and was estimated by Munich Re, the global insurance company, to have caused twenty billion dollars' worth of crop losses. At least eleven billion dollars of that total was paid by taxpayers to cover the insurance liabilities to farmers. Among the fastest-growing categories for government crop indemnities have been extraordinary heat and the inadequate deliveries of water. In California alone from 1993 to 2007, reported a team of Stanford researchers, there were more than five hundred million dollars in crop losses from heat waves, floods, and ill-timed rainstorms in the six counties at the heart of California food growing—San Joaquin, Merced, Kings, Kern, Napa, and Sonoma. Four of those counties are in the Central Valley, representing the bulk of the losses.[10]

While it is difficult to determine precisely how many of these liabilities can be attributed to climate change, the scientific consensus holds that the frequency and intensity of such extreme weather is rising significantly. Severe weather events that inflict more than one billion dollars in damage have risen from an average of two per year in the 1980s to more than ten per year since 2010, according to the National Oceanic and Atmospheric Administration. The damages are due partially to the increasing concentration of the American population in coastal areas prone to flooding and areas of high fire risk, but also to changing weather patterns. Over the past thirty to fifty years, concluded the National Climate Assessment, published in 2013 by the federal government's Global Change Research Program, there have been an unprecedented number of heat waves, severe droughts, and heavy rains, which it said are "primarily due to human activities."

We're already seeing the patterns emerge. Since 1980, the USDA reports there's been a shift toward payouts related to singular "extreme events" rather than from the predictable ebbs and flows of the weather in earlier times. More crop payments are coming.

Growing unease over the volatile weather is sending farmers in unprecedented numbers to seek out increased coverage from the

USDA, according to Jeff Yasui, western division head for the department's Risk Management Agency. "Compared to twenty or thirty years ago, farmers are recognizing that climate change is creating a lot more risk factors in climate events," Yasui told me. The USDA, he said, has responded to farmers' growing uncertainties by issuing an array of new policies that broaden insurance protection. Until the middle part of the last decade, most USDA-backed policies provided coverage of, at best, 55 percent of lost income. Now, Yasui said, far more expensive policies covering up to 85 percent of lost income are getting more and more common. "You've got farmers saying, 'Boy, the weather is really changing. I've got to do something just in case.'"

In another twist heightening the public's cost burden from climate change, the government is increasingly on the hook for more and more of those policies. The USDA permits private insurers to assign up to 20 percent of their riskiest policies directly to the government, reducing their exposure to farms facing the greatest uncertainty. These policies, known as Assigned Risk funds, are increasingly tied to phenomena related to climate change, according to Yasui.

Ironically, the agriculture agency's internal guidelines prevent it from being able to fully plan ahead for the financial impacts from climate change. It's prohibited, said Yasui, from creating insurance rates based on predictions of future conditions. Even as the impacts for farmers accelerate, when the agency readjusts rates every five years it is only permitted to make changes based on weather in the past. Thus, the agency is compelled to confront the growing risks with only one eye open, on catastrophes that have already happened; the other eye, on future projected risk based on those past experiences, must be closed. (Such a position is not as rarefied a way to detour the realities of climate change as one might think; the state of North Carolina passed a law in 2012 prohibiting the use of the words *climate change* by state auditors assessing the insurance risks from floods and extreme weather.)

Private insurers are under no such limitations. David Freidberg is the CEO of a company born of the chaos generated by climate change. Freidberg was a top executive at Google when he started The Climate Corporation in 2009 to offer weather insurance to farmers. He opted out of the federal guarantee program to offer more substantial coverage to farmers that avoids the constraints on rates imposed by too little information. Freidberg conceded that he doesn't know a lot about the details of farming; what he does know are algorithms. He designed a technology that meshes satellite imagery with computer models capable of divining specific weather patterns down to 160-acre-parcel snapshots, identifying the weather conditions right down to individual farms. The policies offer insurance pegged to a specific temperature or weather condition (excess rain, frost, temperature spikes) that could impact future yields. If the event happens, farmers collect. The Climate Corporation's premiums are as much as double those of the federally subsidized system, but farmers are able to obtain damage payouts immediately (rather than waiting to see the financial impacts of adverse weather), and the company offers a level of prediction that is far more precise than the USDA-supported system. The company's own money is on the line, so Freidberg spends a great deal of time devising climate models looking forward as well as back.

"We're bringing a high-tech approach to understanding agronomy and digitizing the relationship between crop yields and the weather and the environment," Freidberg told me in his office in San Francisco. Spread out before us were rows of techies sitting at computer monitors, sporting T-shirts and earrings, who might have been at home at Twitter, just five blocks away. Instead of checking recurring words for advertising purposes, though, they conduct algorithmic analyses of the weather. What he's learned is how past climatic conditions are no longer reliable predictors of future conditions. "California farmers," he said, "are increasingly facing unpredictable yields and unpredictable profits as a result of the increasingly unpredictable snowfall and rainfall in and around the state."

"In most places in the United States," he continued, "we're seeing more volatile and extreme events than we've seen in the historical record, and that's happening with greater frequency. We're seeing that the rainfall patterns aren't simply predicted by the rainfall averages we've seen in the past. We're having more droughts and more major flooding events than we've seen in the historical record and with greater frequency. And so part of our work here is associated with identifying the trends in the unexpected, but also the deviations, the trends in volatility and extreme events. That starts to inform us about what we think the future might hold. . . . As the weather becomes less predictable and more volatile, farmers are going to see more wild swings in their profit. And it's going to start to cost them more to insure against the swings they're experiencing."

At the end of 2013, Friedberg's predictive abilities in a world in which unpredictability is getting more predictable captured the attention of the agribusiness giant Monsanto, which bought The Climate Corporation for nine hundred million dollars. Monsanto is best known for its highly controversial marketing of genetically engineered seeds, introducing genetic material from other species to create inbred resistance to pests, drought, and other environmental conditions. The purchase enables the company both to offer farmers the opportunity to hedge the risks from climate change, and to supply an ostensible solution with its seeds. The Climate Corporation's exclusive and intimate glimpse into the conditions faced by farmers may also be used to help devise seeds adapted to the conditions prompted by climate change. Food, after all, is one of those necessities that will always have a market; it's pretty much irreplaceable, it will sell, to someone, even as the price rises. (Three months after Monsanto's purchase of The Climate Corporation, it purchased the obscure Iowa-based soil analytics company Solum, which applies sophisticated analytic tools to assess the conditions of soils—suggesting that Monsanto sees great potential in mixing weather data with soil data, an invaluable opportunity to exploit the conditions created by climate change.) It is not the purpose of this book to explore the controversy over Monsanto's most famous product, genetically engineered seeds, so let's set that aside.[11] But what Monsanto's purchase of The Climate Corporation makes clear is that there are great profits to be made by those in a position to monitor and predict the increasingly erratic conditions that farmers face.

It also suggests how powerfully the forces unleashed by climate change are altering the landscape for food growing across the country and around the world. Farms will rise or fall based on their ability to keep pace with the tectonic shifts under way. And as new seed varieties are devised in breeding centers across the country—of the gmo and traditionally hybridized varieties—food will be getting more expensive. This is the price we all pay for the impacts of climate change, both at the grocery store and in our support to the nation's crop insurance system, at this stage the major buffer for the consequences of climate change in the nation's fields.

The climate scare that has rippled through California and the nation has led to increasing recognition of the vulnerabilities of the industrial agriculture system to the particular conditions wrought by climate change. Farmers and agronomists are starting to look back to some of the fundamental principles of agriculture before the onset of the chemical era. That era delivered monumental increases in yields over the past half century, but the paradigm is starting to shift as evidence of its costs accumulates. The mass application of chemical fertilizers and pesticides degrades the quality of soils and drastically reduces their ability to generate the organic matter that allows them to absorb water. Repeated growing of the same crop, season in and season out, depletes the earth of minerals that it needs to create fully aggregated soils that, in the long term, are far more resilient to the impacts of extreme weather events. A thirty-year experiment by the Rodale Institute in Pennsylvania, which compared yields of conventional and organic wheat and cornfields—chosen due to those crops' heavy reliance on chemical inputs—showed that over three decades, the latter were far more able to withstand the stresses of drought due to increased water absorption from more organic matter in the soil (it also showed that the yields, and profitability, were roughly equal).[12] Other shorter-term studies have shown similar results, notably in demonstrating that farms with a variety of crops are more capable of withstanding hurricanes and other weather extremes than monocropped plantations, where chemicals deplete the soils' natural defenses.[13] Conversely, the USDA predicts that yields of many crops will decline as the ability of the agricultural system to rebound from environmental stresses diminishes. High-end tree fruits like those San Joaquin Valley cherries will likely be the first to suffer.

As the USDA confronts the specter of rising costs from greenhouse-gas-induced weather changes, the agency has started harking back to ecological principles long promoted by advocates of agricultural reform. The concept of sustainable agriculture seems to have crept into the USDA's lexicon right alongside climate change, sounding notably unlike the factory farm booster it's been in the past. "Adaptation measures," declared the agency's 2013 climate report, such as "diversifying crop rotations, integrating livestock with crop production systems, improving soil quality, minimizing off-farm flow of nutrients and pesticides, and other practices typically associated with sustainable agriculture are actions that may increase the capacity of the agricultural system to minimize the effects of climate change on productivity." Those are powerful words coming from an agency that for years encouraged intensive, industrial-scale agriculture reliant on large quantities of chemical inputs. The USDA's call for composting soils, rotating crops, and decreasing the use of agricultural chemicals sounds very much like what has been on the recipe list of the sustainable farming community for decades now.

Pause for a moment to consider the USDA's use of the word *sustainable*. The word amounts to a claim on the future, suggesting that it's worthwhile taking actions now that will continue to pay off with dividends—in more resilient farms, more food, and more money—in the future. What was the system before? Non-sustainable? The debate over how best to grow food for the long run echoes the debate over redesigning our energy system: Farmers are now being encouraged by the nation's premier agricultural agency to reformulate their food growing in a way to better ensure their ability to continue doing so in the long term. The soil is renewable—meaning it operates in cycles of replenishment—just as energy sources can be renewable.

Similarly, the California Climate and Agriculture Network has been advocating that farmers be encouraged with subsidies or carbon credits to engage in practices that enhance the ability of the soil to actually sequester carbon—for instance, by tilling the soil less frequently or by churning biomass residue back into the soil. "Farmers could actually play a significant role in reducing our greenhouse gas load," commented the group's director, Renata Brillinger.

Reading the USDA's recognition of principles that have been central to the family farm and organic movement, I couldn't help but think of a counterpart—health care, for humans. The crop insurance program is like a subsidized health care program for plants, and for the farmers who grow them. In countries that have nationally subsidized health care, governments have found that they have a serious financial incentive to keep people healthy; if not, the government foots the bill. This is one of the reasons why European governments have been trying to address toxic chemicals more aggressively than we in the United States: If their citizens show up with symptoms of exposure, they pay the cost.[14] The same with crops: The less healthy and productive our farms, the more the government pays. The more resilient the food system is, the less, in the long run, it will cost the government to prop it up.

The crop insurance system could be a powerful leverage to steer farmers toward practices that encourage resiliency. It could leverage premiums to discourage practices like overtilling or excessive application of chemicals that diminish the health of the soil. But the government seems to be moving in contradictory crosscurrents. The

GAO set off the alarm of burgeoning costs from climate change's impacts on American farms. The USDA identified legions of impacts from climate change, from the debilitating effects of temperature rises on crops to the erosion of soils by violent rainstorms.

Then came the Farm Bill: After three years of contentious debate, in February 2014 President Obama signed a new bill governing the multibillion-dollar subsidy and insurance system for farmers. Most notably, it eliminated six billion dollars in direct payments to commodity growers and dramatically expanded the crop insurance system to ninety billion dollars over the next ten years. That's likely an understated figure, given the fact that farmers received eleven billion from the 2012 drought alone. Equally significant, it expands the revenue protection part of insurance to cover up to 85 percent of a farmer's revenue losses (up from the previous 75 percent)—creating extra incentives to plow up low-yield marginal lands, including grasslands and wetlands, prime locations for numerous species as well as key to soaking up water for long-term storage and flood control. It also sent a mixed message with its financial assistance program, increasing by some $30 million (to $52.5 million) a program to help conventional farmers switch to organic growing, while eliminating a $5 billion program to support soil conservation efforts—thus drastically reducing what is probably one of the most potent ways to help create greater climate resiliency. So on the one hand, the USDA is attempting to alert the public about the climate-related stresses that loom for American farmers, and on the other its crop insurance and subsidy system does little overall to alter farm practices—overtilling, mass application of chemicals, monocrop farms—that undermine the land's ability to respond to the pressures caused by climate change.

Crop insurance, finally, is one of the clearest examples there is of the government (meaning us) assuming the burden created by our use of fossil fuels. And what about the other part of the Mobius strip, in which farmers contribute to the plume of greenhouse gases that rise from the fields into the atmosphere, and that then come down to bite them in the growing of our food?

Identifying the carbon footprint of food is a complex enterprise. Unlike, say, airplane emissions, which come from a clearly defined single source over a clearly measurable period of time, food is the ultimate step in a multiple-stage life cycle, each of which has different greenhouse gas implications. A seed is planted: Was it born in a greenhouse, requiring heat, or in a field? The field is watered: Is it irrigated, involving electrical pumps, or rain-fed? The field is protected from pests: Is it treated with synthesized chemicals or organized around buffer crops that repel pests? The crop is harvested: by mechanical harvester, or by farm laborers? Finally, it's delivered to you, to your neighborhood supermarket or health food store: by truck, airplane, or ship? Every decision has greenhouse gas consequences.

A quick look at a couple of examples gives us a sense of the major greenhouse gas contributors at every stage of the food distribution system. Let's start low on the food chain, with the humble potato chip. In 2011, the British processed food company Walkers decided to reduce its carbon footprint, and asked the Carbon Trust to assess the contribution of its chips—"crisps" in British English—to climate change. It committed to publishing the result—which it does on every packet of Walkers you buy in the UK. What it discovered was that for each four-ounce package of its most popular flavored crisps—cheese and onion—eighty grams of CO_2 were released into the atmosphere. The breakdown reveals how every stage of every product is loaded with contributions to climate change—from the growing of the potatoes and spices (36 percent of the total) to processing them into slender tasty crisps (17 percent) to packaging them into nifty four-ounce packets (34 percent) to transporting those bags to retail outlets across the country (10 percent) to disposal (3 percent).

And now let's take a step up, to an entire supermarket. In that same year, the British government requested that major retailers voluntarily offer labels identifying the energy it took, and the emissions along the way, to get food from the field to grocery store shelves. The country's biggest chain, TESCO, took them up on the offer. By the end of 2012 the company, with more than five hundred stores in the UK, published on their website a chronicle of the carbon footprint of more than a thousand products. There is the 170 grams of CO_2e (a scientific measurement referring to CO_2 and its "e"quivalents—greenhouse gases) from a kilogram of Welsh New Loose Potatoes; the 110 grams of CO_2e for each 100 grams of TESCO Mushy Peas; and the 360 grams per 75-gram serving of TESCO's Finest Egg Fettuccine. The list goes on, from vegetables, fruits, chicken, and steak to physical items like Tupperware and dishwashing soap. It's based on a life-cycle analysis, compiling emissions at every stage of food from the field to the trash can, the growing, processing, packaging, distribution, retail preparation (refrigeration, for example), use by customers, and disposal. TESCO's list,[15] a long litany of just about every product you'd encounter in a supermarket, is probably the most comprehensive list yet compiled about the particular carbon footprint of the food we eat. Altogether, those numbers read like a ticker tape of how our very act of staying alive and consuming food contributes to the atmospheric chaos. Needless to say, this is not a plug for TESCO; they offer an unusually specific list that can easily be extrapolated to the products of any other conventional supermarket.

There has been no comparably comprehensive effort in the United States to assess the footprint product by product. The Environmental Working Group published a broad assessment of food groups' carbon footprints.[16] Based on meal-sized portions, for example, they concluded that green vegetables have the smallest footprint, the equivalent of driving about half a mile. Up at the top of the footprint scale were pork (about 2.5 miles), cheese (3 miles), beef (more than 6 miles), and lamb (off the chart at over 7 miles).

This is of course a broad-brush approach, and ultimately varies with the distance the food travels from farm to fork and the amount of processing it goes through along the way.

Several conclusions can be drawn from these and other findings. First, processed food is far more greenhouse-gas-intensive than fresh food, due to its processing costs, transport, and packaging. Second, distance traveled is a key factor. Imported tomatoes or any other food will have a far larger footprint than locally grown food. Third, organic food does not *necessarily* offer a smaller carbon footprint. Yields per acre—per ton of greenhouse gases—are sometimes (though not always) lower on organic farms, meaning that in some instances it takes more land to produce the same amount of food grown on a conventional farm, upping the individual footprint from unavoidable greenhouse gases (some tilling, transport, processing). But of course, organic farming comes with numerous other benefits—hardier and healthier varieties, fewer chemical toxins, and farming practices that return far more nutrients to the soil. This soil environment rich with organic matter is not only more resilient to drought, but also highly effective at absorbing and storing greenhouse gases. Fourth, intensively raised red meat is pretty much the worst from a climate perspective.

And so we return to the question of how, ultimately, we integrate the costs of the carbon footprint of food. A group of scientists at UC Berkeley estimated that each thousand dollars spent by Americans on conventionally grown food releases another ton of greenhouse gases into the atmosphere. Adding 1 percent to the price of food, they estimate, would begin to mitigate the costs from climate disruption.[17] Another group of scientists suggested in the journal *Nature Climate Change* that meat consumption should be penalized in order to reduce the massive contribution of livestock to our methane load.[18] A "tax or emission trading scheme on livestock," they argued, "could be an economically sound policy that would modify consumer prices and affect consumption patterns."

In the end, as climate impacts increase, the web of costs associated with cultivating plants and animals for food grows ever more complex—from research into new varieties to increased input costs to government payouts to increased sticker prices for consumers. Cultivated agriculture, after all, involves the intervention of humans at every stage.

But when it comes to climate mitigation, even the uncultivated gets complicated. In our next chapter, we will explore trees, whose critical role in storing CO_2 makes them valuable not just for the food that may dangle from their branches, but for the very act of being alive. Trees in tropical forests just need humans to get out of the way. They're one of the most potent weapons against climate change we have. Yet the one move that could reduce the impact of all our carbon footprints has triggered some of the fiercest controversies over how to do just one simple thing, which is nothing at all. Keeping a tree standing, and determining the price for that simple act, turns out to be a minefield in which financial interests collide with new frontiers of science and technology.

CHAPTER 4

The Forest for Its Carbon

The Tree

A century ago, if I'd made the long journey to the Brazilian port of Manaus on the Rio Negro, one of the great tributaries of the Amazon River, it would most likely have been in search of rubber. From here, American and European adventurers and schemers ventured into the jungle in search of trees and their sap. Hundreds of millions of dollars flowed through Manaus, the capital of Brazil's largest state, Amazonas. Today the city still feels improvised, as if its inhabitants were accidentally dropped into the middle of the jungle and told to build.

Manaus is hot, it teems with insects, and red and purple flowers grow through every sidewalk; even subdued, the jungle seems to burst at every turn through the urban surface. It's a boisterous sprawl of two million people with traffic-jammed boulevards, street vendors hawking sugarcane juices squeezed on the spot, and a port where colorfully painted wooden freighters, loaded with goods and people, jostle with simple canoes in the water. Manaus is the last stand of "civilization"; from here, in every direction, sprawls the Amazon.

In a plaza of gnarled trees and blossoming vines sits the nineteenth-century theater built by the rubber barons to feature their favorite performers from Europe. The Teatro d'Amazonas is a surreal testament to a time when those barons came to the forest to export its riches and, in the process, imported the titans of opera from home to entertain them. Also known as the Manaus Opera House, the theater is a spectacle of absurdist grandeur—an opulent marble palace in the jungle, still with its original tropical hardwood seats and a hundred years of congealed tropical moisture lingering heavily in the air. Above the orchestra seats, fleur-de-lis imprints bear the names of the great European cultural figures of the nineteenth century—Goethe, Verdi, Beethoven, Hugo, and others. Fans of the filmmaker Werner Herzog may remember this as the place where the crazed Fitzcarraldo—in the film of the same name—stopped while on a frenzied quest to build his own opera house in the jungle. Though it now features modern-day performers, the theater seems stuck in time, a time when foreigners came to get a taste of the "high" culture they left behind while the natives toiled on their behalf in the jungle.

A century later, there's a new generation from Europe and America who have been embarking from Manaus in search of the jungle's bounty. Only this time many of them are not looking for trees to cut down or for sap to extract; they're looking instead for what the jungle *is*—the world's largest sink of carbon dioxide.

These foreigners are not interested in exporting from the jungle, but rather are seeking to preserve it exactly as it is. They're hunting for a new commodity—the CO_2 embedded in Brazil's billions of trees. They are seeking a living ecosystem in exchange for the pollution that fossil-fuel-intensive industries send into the atmosphere. The trees here have not been cultivated, like those California cherry groves; they arise out of the churning jumble of nutrients in the soil of the jungle.

Trees are our partners in respiration: They inhale carbon dioxide and exhale oxygen, and we, of course, do the opposite. Foresters estimate that roughly half of every tree (the exact figure depends on species, elevation, and other factors) is made up of carbon dioxide—just as we humans are composed significantly of oxygen. The CO_2 is embedded in the trunk, the leaves, the bark, the very essence of the tree. When a tree is cut or burned, that CO_2 is released into the atmosphere. Equally important is the fact that the very process of clearing or cutting trees involves widespread destruction of the forest ecosystem; half of the CO_2 in a forest is in the soil and ground vegetation. In the same way that tilling soil sets loose carbon dioxide long sequestered in the earth, felling trees causes serious collateral damage. The United Nations estimates that deforestation accounts for 15 to 20 percent of all greenhouse gas emissions.

For a glimpse into the interplay between trees and greenhouse gases, take the wooden desk on which this book is being written. I don't know where it came from, but it is made from a birch tree, which means it most likely came from the middle latitudes of the United States or Canada, and probably from a second- or third-growth forest, given how few virgin forests are left. The Congressional Research Service has helpfully calculated how much CO_2 is sequestered in a growing North American temperate forest, the likely source of wood for my desk and for much bigger things, too, like possibly your house. A twenty-five-year-old American birch forest sequesters about 1,760 pounds of CO_2 per acre per year, an average of 2.5 pounds of CO_2 per tree. (For a 120-year-old forest, by contrast, the sequestration rates are more than double that, because trees and the surrounding ecosystem had time to grow in complexity and size.)[1] The USDA Forest Service estimates that the total amount of carbon stored in American forest ecosystems amounts to about twenty-five years' worth of the total US greenhouse gas emissions in 2012. Wherever its precise genesis, then, there's no dodging the fact that trees felled from the forest are significant contributors to the atmospheric burden of greenhouse gases.

Some of the carbon remains, it must be said, in the deadwood of my desk. Just over half of a tree's CO_2 remains in a piece of wooden furniture after ten years, according to the USDA. Which means my desk is something of a carbon sink, for a while, anyway; that rate declines each year as the wood decays and slowly releases what was once its life-giving CO_2, and after a century little more than

0.003 percent of its original CO_2 remains.[2] In other words, turning a tree into a desk just slows down the release of carbon dioxide, it doesn't prevent it. Needless to say, the ecosystem my desk supports is hardly that of a living tree—it sustains little more than myself, my computer, and the papers that are sprawled across its top. (This of course would be laughable were it not for the fact that the country's largest timber company, Sierra Pacific Industries, successfully demanded that the CO_2 sequestered in its "wood products" be counted by the state of California as saved carbon when tallying up the company's greenhouse gas emissions.[3])

By far, the greatest abundance and diversity of trees are in the tropics. This narrow band around the equator harbors species-rich ecosystems, named Vavilov Centers after the Russian botanist who figured out that the equatorial belt is the source pool for 90 percent of the planet's biodiversity, and thus critical to the planet's ecological health. Not only do tropical forests sequester more carbon than trees in other regions—at least a quarter more than the birch trees of North America—but they do so year-round and contribute to a dynamic ecosystem that sustains an infinite variety of plant and animal species. Tropical forests offer many other benefits as well: They provide stability to the soil, are great at absorbing water, and are central to the cycle that returns water to the atmosphere after it has fallen as rain. The complex web of life that emanates around a tree—from the insects to the birds to the tree-jumping primates, the lianas that curl up its trunk, and the fungi that live in its shade—is far more resilient, scientists now know, to the shifts in temperature, precipitation, and other factors triggered by climate change. Even these descriptions, though, can't capture fully the magnitude of forests' role in the global ecosystem: They are also referred to as the "lungs of the earth," key to that life-giving oxygen–CO_2 respiratory exchange with all the fauna of the planet.

Thus, the deforestation that has been rampant for the past decades has immense consequence for our ecological equilibrium, particularly as the effects of climate change accelerate. Deforestation accounts for more greenhouse gases than car, air, and ship travel combined. More than thirty million acres of tropical rain forest—as much land as the state of Mississippi—are destroyed every year, according to the United Nations. That burning and cutting of forests, according to an EPA assessment, leads to the release of about 1.5 billion tons of CO_2 into the atmosphere.

This makes standing forests one of the most valuable tools for combating climate change, because living trees continue to suck CO_2 out of the atmosphere. They reduce greenhouse gases by the simple virtue of being alive and growing. The precious role of trees in our planet's ecosystem, and in particular the central role they play in the sequestering of CO_2, has given rise to an entirely new legal concept: carbon rights. Just as mining companies can buy the rights to the minerals beneath your land but not the land itself, carbon trading companies are buying the rights to the carbon in trees but not the tree itself.

For polluting companies it can be a lot cheaper to purchase the carbon rights and allow all those trees to continue sequestering CO_2 than it is to reduce the quantity of greenhouse gases they're emitting back home. It's a transaction that ties those of us in the developed world to the forests in the developing world—and the people who live in them—like never before.

But trees have traditionally been valued lying down, as lumber, rather than standing up. Or, even worse, they are seen as obstacles and just burned out of the way to make room for farms or housing. So how valuable are they? What does it take to keep them standing?

Of all the equatorial Vavilov Centers of biodiversity—which also include vast rain forests in countries like Papua New Guinea, Indonesia, Malaysia, the Congo, and Colombia—Brazil's forests are by far the largest. So it was here, in the richest place for trees on earth, a place that is central to the long-term fate of the planet, that I would try to discover what it means to put a price on one tree standing.

Opera stars were once imported to bring high culture to the rain forest; now the rain forest itself has been elevated into the highest realms of reverence, the high culture for our era of climate change. Over the course of a week in Manaus, I spent time with one of the people who have been attempting to navigate the line between treating trees as an extracted commodity and treating them as the earth's global environmental treasure.[4] Eduardo Braga ran Amazonas, four times the size of Texas, with a populist swagger as its governor from 2003 to 2011, and then became the region's senator in the national congress. Braga built his rapid political rise on a two-stroke approach to economic development in the state. He lured global industries into Manaus's free trade zone, including a new factory making Harley-Davidson motorcycles situated on the outskirts of the city. And he tried to devise ways of making preservation of the jungle profitable, such as establishing a green Free Trade Zone, now a center for a new round of exports ranging from Brazil nuts to the health drink acai, to perfumes and "natural" cosmetic ingredients. In 2013, he received a Global Conservation Hero Award from Conservation International.

The jungle itself still has a hold on him. Braga rhapsodized about the Amazon's "magic" when I interviewed him in his governor's office the day after a torrential rainstorm: "You can feel it," he said. "The forest has a special energy, but at the same time it's very quiet. If you are walking during the night in the forest, it's amazing because the silence is broken for a small noise that you can hear, a small movement seems like a big scream in your face. In the other moment, if you are there during a winter storm, like we had yesterday afternoon, it comes like a fire in the forest. It's a special chemistry we have with nature."

That chemistry is tested every day by the financial realities in one of Brazil's poorest states. A day later, when I joined the governor and his entourage at an evening public meeting, Braga was not quite as poetic. In an open-air amphitheater the size of a basketball court, the scene was festive, with beer for sale in plastic cups, filled straight from the barrel. By nine o'clock that summer night, the breeze was finally blowing to cool the sweltering heat. From the side of the stage I watched the flamboyant Braga's white shirt turn wet with sweat as he exhorted the crowds: "These are our forests, and no one can come here and take them. They are our forests, and they must pay if they want to keep them!" He got roaring applause for that line.

The "they" in this case was "us"—meaning those who do not live in the Amazon, namely Americans and Europeans who have come to recognize the central role played by forests in the earth's ecology. If the forest is so important for us—for the planet—then its time to pay for it. Braga suggested I visit an offset project he'd helped establish south of the city.

It took a two-hour flight in a six-seat Embraer to get there. From Manaus we flew over what looked like endless fields of enormous dark green broccoli—the rain forest. We flew low enough to also see the long ribbons of brown running through them, and the empty patches, like fallow fields—ominous signs of logging.

We landed on a short dirt airstrip carved in a perfect rectangle out of the trees in the Juma Sustainable Development Reserve—a tiny settlement within the 2.5 million hectares he'd established as a protected zone in an area surrounded by illegal logging. I walked on a well-maintained path past Brazil nut and date trees, several of which had young men scurrying up their trunks with harvesting bags slung over their shoulders. At a small community center, topped by palm reeds, a stove in a small kitchen was bubbling with vats of acai collected from the berries that dangled seemingly everywhere. A botanist consulting with the community brought me into their on-site nursery to smell a flower that he said provided the basis for the French perfume Chanel No. 5. I can still remember that sensation, the whiff of an aroma from the elegant quarters of Paris there in the middle of the Amazon.

I'd been told by Braga, as well as environmental NGOs like the Environmental Defense Fund, that Juma was the model offset. The sponsor was the Marriott International hotel chain, which provided two million dollars to build the infrastructure—including a small community meetinghouse and help with marketing the community's forest products—and promotes the project as a way to offset its guests' carbon footprint. I recalled the Marriotts I've stayed in over the years, the laundering of the sheets and towels, the whirring of hot or cold air through the vents, the flicking of a million light switches. It seemed vaguely reassuring and also unnerving that all those greenhouse gases I and Marriott's millions of other hotel guests generate could find their equivalent way out here in the middle of the Amazon. It seemed as if this little pocket of the jungle had been given a shot at some form of self-sufficiency that does not involve cutting down the trees. But even that comes with a caveat.

Braga had created a program, called *Bolsa Floresta*, which provides a monthly stipend of about thirty-five dollars to families in this and other protected areas as an inducement to resist the temptation to go to work for one of the logging companies, which were encroaching on the area from the north. The money is provided by regular contributions direct to bank accounts of the community's women (who are perceived as being more financially responsible and accountable than their husbands). And how far does that sum go in the jungle? It costs almost twenty dollars to travel both ways, six hours each way by canoe, to get to the nearest bank. Several people complained that it wasn't nearly enough to compensate for the food growing they used to do on now protected lands.

An assessment by the British Institute of Development Studies concluded that while the Juma project offers valuable payments for ecological products, and provides some buffer to the lure of black-market work for illegal land clearing, the conservation benefits of preserving carbon in the trees were disproportionately prioritized over economic development. "The development benefits are often secondary to the conservation ones," the institute concluded. That tension between preserving the tropical forests for their carbon and using them for their immediate economic value is one of the key obstacles as the world stumbles toward a hedge against continuously rising greenhouse gases.

Indeed, finding a way to preserve tropical rain forests for carbon offsets has become one of the central challenges facing Brazil, other tropical countries, and climate negotiators, who are trying to conjure money for trees to do nothing more than exist. At the 2009 climate negotiations in Copenhagen, the United States, Britain, Norway, and other developed countries committed $4.5 billion to launch a global initiative that would begin to assess the value of the world's tropical rain forests. Four years later, in 2013, at the Warsaw Climate Change Conference, further rules set criteria for tropical countries to meet in order to receive payments in return for reducing deforestation or launching sustainable forest management strategies. Now global efforts focus on what the forests are worth, and who will pay to keep them standing.

The first step has been to inventory the world's forestlands, identifying which ones are still intact and how much carbon they contain. In a NASA station near Menlo Park, about thirty miles south of San Francisco, the fruits of the Obama administration's 2009 Copenhagen commitment can be seen. There, at the Ames Research Center, scientists monitor images from satellites launched by the United States, which along with other NASA duties provide ongoing pictures of distress in the world's forests.

I sat with Christopher Potter, head of satellite imagery at the station, as we took a round-the-world tour in front of a bank of computer screens. From ten miles high, the satellite registers the heat generated by the world's fauna—a spectrum of browns and yellows and greens set against the outlines of countries, continents, or even individual forests. With Potter at the controls, we zoomed in on the forests in the

Malaysian side of Borneo, where we saw deep greens giving way to pale greens giving way to brown—indicating forests that had been cut down. And then there were the perfectly formed rectangles of green, a sure sign of the palm oil plantations that are rapidly replacing the once virgin tropical forests. From where we sat, the earth looked like an organism throbbing with the energy generated by photosynthesis, and with wounding gashes everywhere we looked.

You could see the forests that were thriving, the forests at risk, and the forests that are already lost. Faced with this precise geography of threatened forests, you could be forgiven for thinking that saving them would be relatively simple. But the effort to conserve forests through payments, whether government or private, has been repeatedly confounded by the reality on the ground, a complex root system of economics, science, entirely new legal concepts, and a dash of firepower.

"Ten years ago, no one thought of carbon as property, something people could own," said David Takacs, a professor at UC Hastings College of the Law who has written extensively on the legal challenges presented by a United Nations program called Reducing Emissions from Deforestation and Forest Degradation. Also known as REDD, the program aims to help developing nations assess and protect forests as carbon offsets. "Now we're trying to think of carbon as a property," he adds, "But who owns the land? Who owns the trees? Who owns the right to the carbon credits? Those are big questions."

That's putting it mildly. The matter of who owns the right to offer the earth's trees to far-off polluters is one that strikes directly at the equity debate that lies at the heart of the climate conundrum. Should forest dwellers be asked to forgo their subsistence use of the forest so a greenhouse gas emitter in the United States or elsewhere can emit more greenhouse gases?

On almost every continent, indigenous associations have come out strongly against REDD because it can result in the people themselves—with the smallest carbon footprints—being excluded from land they have long relied on by companies with the largest footprints. Numerous studies have shown that the best conserved forests are those that have been home to indigenous peoples, who protect the forest because they live in it. But the phenomenon of displacement has occurred numerous times, in different places with different players but with a similar denouement—notably, between 2010 and 2013 in Kenya, Tanzania, Papua New Guinea, Peru, and Indonesia. Those are just reported incidents; getting journalists or independent monitors into remote forest locations can be difficult. In 2006, the enviromental journalist Mark Dowie's seminal book, *Conservation Refugees*, outlined how indigenous people were being excluded from their homes in forests because conservation groups wanted to preserve the "wild," pristine essence of the forest, which has been interpreted to mean "without people." Now that phenomenon is being repeated, with a twist—offset refugees.

The initial REDD concept relied heavily on commitments from private companies to use forests as emission offsets. But the resistance by forest dwellers and the unwillingness of most of the world's carbon markets to permit the use of those offsets prompted the creation of a more evolved version of the plan, called REDD+, which rewards the governments of tropical forest countries with financial payments if they reduce their rates of deforestation and offer alternative economic opportunities for forest dwellers. The expanded approach largely shifts the financial burden from private companies to developed country governments and international aid agencies like the World Bank.

But the closer you get to the actual forest the more difficult it becomes to navigate between the fundamental questions of equity that continue to lie at the heart of the very concept of REDD and REDD+: Should tropical countries be expected to forgo the immediate financial benefits from exploiting the forests' resources on behalf of the rest of us on the planet, for whom they mean little financially but are monumental in their ecological importance?

If I'd made the journey sixty years ago to Brasilia, the Brazilian capital, it would most likely have been in search of meat. Back then, Brasilia was a small trading outpost in the middle of Brazil's broad swatch of tropical savanna in the center of the country. Cattle were bred across the broad hilly *cerrado*. Then the government decided to move the capital from the coast, in Rio de Janeiro, to the plains of the interior, and by 1960 construction was complete. Designed according to the most *avant* ideas of the time by the architect Oscar Niemeier, Brasilia, even now with a population of two million plus, still retains its futuristic tableaux—a geometry of four grids, with one district each for commerce, residences, recreation, and the government, a complex of grassy walkways and 1960s-era boxy buildings flanked by broad boulevards.

Off one of those boulevards, across the street from a lush little pocket park, are the offices of IBAMA, Brazil's environmental protection agency. A colorful mural on the arch over the entryway conjures Chico Mendes—whose efforts to unionize contemporary rubber workers helped lead to his assassination in 1990. He's now seen as a hero of the country's labor and environmental movements. The rubber barons left testaments to their success in exploiting the jungle's productive largesse with places like the Manaus Opera House. Now there is a testament, in Brazil's premier environmental body, to the man who challenged the efforts of the rubber barons' descendants to maintain that control into the twentieth century.

The cutting and burning of trees has been rampant in Brazil over the past thirty years—with ranchers and illegal foresters routinely burning massive swaths of rain forest, either to clear them for cattle grazing or soybean fields, or to log the precious hardwoods that are getting more expensive as they dwindle in number. The government has expressed alarm at symptoms of climate change already breaking out in different parts of the country—unexpected droughts in the Amazon and increasingly violent storms that threaten the already fragile infrastructure in cities like Rio de Janeiro and Salvador.

Brazil has pledged to reduce its emissions from 1990 levels by 39 percent by 2020. Since reducing rates of deforestation by 80 percent is a central ingredient to their strategy, the government launched an intensive effort to stanch the destruction of its tropical forestlands. New laws require landowners in the Amazon to preserve 80 percent of their forest property in pristine form. The government also devised a blacklist of Brazilian retailers caught purchasing goods from anyone illegally logging or burning off forestlands. The laws and blacklist have had an effect: The rate of deforestation in Brazil has dropped significantly. Between 1996 and 2005, an average of 19,500 square kilometers were cleared annually; by 2010, that rate had dropped to about 7,000 square kilometers, and in 2012 it was down to a little over 2,000 square kilometers. The agency has levied more than a billion dollars in environmental fines since 2009. By 2013, deforestation was no longer the leading source of greenhouse gas emissions in Brazil; it had dropped to third place after transport and industrial sources.

Luis Evangelista, the chief of IBAMA's enforcement unit, ushered me onto a floor filled with glowing computers—the center of IBAMA's high-tech battle against deforestation. He left me with a pair of twenty-somethings with master's degrees in environmental studies on a floor populated entirely, it seemed, with young men and women in jeans. They could leave the building for a rock concert and no one would have to change clothes. Here they sat, half a dozen of them, focused intensely on computer screens displaying satellite images from the jungle. We stopped at the work station of Werner Goncalves, a forestry engineer in a yellow T-shirt, sporting a goatee. On his screen was a palette of pixels in deep purple and splotches of hot pink, photographs taken just days before from a satellite orbiting high above the Amazon.

The satellite pictures were really snapshots of heat, in twenty-five-kilometer parcels. One way to look at the forest is as a huge engine of photosynthesis: Where there's heat, there are trees; that's the deep purple on Goncalves's screen. And where there is no heat, those trees are gone. Goncalves gets as many as two hundred of these snapshots of the jungle every fifteen days, each of them a peek into whether the status quo is holding or disruptions are taking place on the ground. He pointed to the pink splotches on his screen: "That's where someone is attacking the forest."

Goncalves and each of the computer-bound satellite analysts send those images off to police in Brazil's six states that still contain substantial rain forests. The police cross the images with a GPS map to guide them to sites of deforestation.

I joined a police force in Pará state, neighboring Amazonas, where we spent three days on the hunt for illegal loggers. We drove at high speeds down dirt roads in a three-car convoy loaded with heavily armed state police tracking satellite images sent to our navigator, a policewoman in the lead SUV. Our satellite-led mobile stakeout was centered on the town of Tailandia, ground zero for the illicit business and the powerful interests that support them. Three months earlier, there'd been a gunfight in Tailandia between these same police and the loggers.

We followed those satellite images down barely discernible dirt roads, through gullies and streams and then into the grasslands that take over after the forest has been destroyed. Rolling onward, we crested a hill and came upon a diorama from hell. Below us were what looked like half a dozen brick igloos pouring acrid black smoke into the air. The "igloos" were kilns, burning trees to make charcoal. Ahead, stretching into the distance, were acres of cleared land, the tree stumps standing like specters. The scene was heart-stopping: The Amazon had been cut into pieces and fed into primitive ovens. Fires raged inside. All around each oven were carefully piled black bricks of charcoal.

The police approached a terrified man in tattered shorts and a T-shirt, wearing simple worn leather sandals. He looked to be in his fifties, though he was probably far younger, and lived in a lonely, ramshackle cottage in the distance. His teenage daughter was suckling an infant. The leader of our squadron, a colonel with a rifle slung over his shoulder, approached him. The man explained, nervous in the face of all this firepower, that he would be selling the charcoal to a middleman, who came by every week. And who was the middleman? The man claimed not to know. But our squad leader said later that day that the charcoal was probably destined for one of the steel mills in an industrial district about a hundred miles away, which is one of the key markets for the cheap, and illegal, fuel derived from the burning of the Amazon.

If the colonel could identify the buyer, he'd add him or her to the blacklist; but, he said, many operate under "phantom companies," and identifying the precise buyer is difficult. The man making the charcoal probably earned about $150 a month out of this grimy, hellish business. He was working on the bottom of the deforestation chain. At the top was the harvesting of the precious hardwoods, which can be sold for as much as one thousand dollars a tree. Then there are the secondary, less valuable trees taken for processed wood and everyday lumber. And then come the remnants of the forest, left behind to be burned into charcoal.

To see the rain forest cleared of trees is one of those scenes you never forget: stumps, tossed detritus, barrenness where there was once lush chaos. And of course, the land cleared of trees loses not only the central ingredient to its ecosystem, and the multiple thousands of species and processes that are sustained in a tropical forest, but also the ability to sequester carbon. We were looking, said the colonel, at roughly three acres of recently cleared land. At 109 tons of carbon per acre in tropical rain forests (an estimate from the US Congressional Research Service that includes trees, roots, and all the other plant life of the forest[5]), another 327 tons of CO_2 had been let into the atmosphere by this hellish scene.

The complexities are also sobering: As the rest of the squad took hatchets and shovels to the kilns, hacking one after another into piles of rubble, the man and his family were ousted from their small cabin. They were sent on their way, to find work, somewhere, out there, down a dirt road that was at least a two-hour walk from the nearest highway. Who knows where? The state offers a stipend for displaced workers, but first you have to get to the nearest city to sign up, and it's rarely enough to support a family. It wouldn't take long, the colonel said, for this guy to be out there again, trying to scrape a living out of the jungle.

And this was in the country that is, by far, the most sophisticated of all equatorial countries in dealing with its tropical forests. Other heavily forested countries, like the Democratic Republic of the Congo and Papua New Guinea, have barely functioning legal systems, though they, too, have been on the target list of carbon project developers looking to offer offsets to companies seeking to burnish their image on preservation projects far from their actual source of pollution.

Getting into the nitty-gritty of Brazil's multifaceted approach to its rain forest brings us into the country's complex and sometimes contradictory balancing act between development and preservation. Valuing the carbon content of a tree is one of the most sensitive and difficult challenges for those trying to clarify the difference between a tree's usefulness now and the almost divine services it provides to the planet.

Guaraqueçaba is located in Brazil's second great ecological treasure, the Atlantic Forest. Far smaller than the Amazon—and thus often neglected by scientists and politicians—it once sprawled across millions of acres of southeastern Brazil. Now, after decades of urban settlement, agriculture, and logging, just 7 percent of the forest remains. What's left of its spectacular biological diversity and sheer chaotic splendor rivals that of the Amazon.

One July afternoon, I found myself standing in that fifty-thousand-acre Guaraqueçaba reserve in the shadow of tree #129, about a hundred miles south of the nearest city, Curitiba. Here I would discover one way to calculate the value of the trees that remain after decades of forest clearing and cutting.

To get to tree #129, I'd followed Ricardo de Britez, a Brazilian forester who works for a Brazilian conservation organization, the Society for Wildlife Research and Environmental Education (SPVS), which in turn works with the US group The Nature Conservancy, which helped establish the reserve. De Britez was on his bimonthly foray to measure the forest. There are 190 stations—stands of trees with small numbered plaques hanging from their trunks—scattered around the reserve, where de Britez and his team routinely measure a random ongoing sample of trees for their carbon content. The goal: that the trees grow, and thus capture more CO_2 with every new inch.

En route, we passed clusters of yellow and white orchids, stepped over the footprints of an ocelot, kept an eye out for the endangered golden lion tamarin, and were bitten by, it seemed, every one of the thousands of insect species native to the area. Finally, we arrived at a cluster of guaracica trees, notable for their white bark, and there it was, with a small metal plaque hanging from its trunk—tree #129.

De Britez took a tape measure hanging from his belt and, with an assistant, unspooled it around the trunk. He declared the tree to be three feet in diameter, with a crown, way up above our heads, about forty-five feet high, and noted the figures in his clipboard. He told me that, extrapolating from its bulk, #129 probably contained about ninety-five kilograms of carbon. That's just under one-tenth of a ton.

Taking the Congressional Research Service estimate that tropical forests generally contain 109 tons of CO_2 per acre, that would give this fifty-thousand-acre reserve an existing carbon sequestration of about 5 million tons. Cut them down or burn them and most of that is released. But let them grow and slowly, every year, more CO_2 is incrementally embedded in the trees and other forest fauna.

At the time I visited Guaraqueçaba, the price of carbon on the US voluntary market was ten dollars a ton. Which means that tree #129, were it to be sold to you or me, would be worth, say, a little under one dollar. Except in this case those carbon rights are already owned—by the American car company General Motors.

The reserve is protected from development by an unusual deal made with polluters back in the United States. In return for agreeing not to cut down the trees, three US companies with huge carbon footprints—General Motors, American Electric Power, and the Chevron Corporation—agreed to invest eighteen million dollars here to preserve the trees as an early test of a REDD project in 2003, back when they thought the United States would soon have a carbon market. In return, they obtained the rights to the carbon in the trees—a concept known as avoided deforestation, akin to buying the rights to minerals under the earth but not the land over them. The companies agreed not to cut down the trees and claimed that the CO_2 accumulating in them as they grow would offset some of their greenhouse gas emissions back home. How much carbon, how many trees, how much money? The three companies, working in cooperation with The Nature Conservancy and its Brazilian ally, SPVS, financed the effort to find out.[6]

So one way to look at tree #129 is that it's worth ninety-five cents on the voluntary market.

Another way to look at that single guaracica tree is that its very existence might cover perhaps one week of average gasoline emissions from driving in one of GM's fabled Hummers. It would take about fifty trees to offset the average emissions of one Hummer for a year, according to the website Carbon Tracker.

Critics, including many in Brazil, have questioned whether the existence of such projects actually serves to discourage reducing emissions, since by transferring the reductions elsewhere it reduces incentives for developing technologies that could have more lasting impact. José Miguez, one of Brazil's chief climate negotiators, sees it as a way for American companies to dodge their responsibility to reduce emissions in their own backyard. "The forest is there!" he told me. "You can't guarantee it will absorb extra carbon. The General Motors plan gives a false image to the public in the United States.

For us, they are pretending to combat climate change." Supporters such as The Nature Conservancy assert that such offsets are the only way to channel the vast amount of capital needed to lower the rates of deforestation.

And another way to look at tree #129 would be from the point of view of the residents of Guaraqueçaba, who have lived for generations in a forest that is now within the borders of the reserve—and have been cut off from the forest's resources since the reserve was established in 2003. I spoke to numerous people in settlements scattered deep in the jungle throughout Guaraqueçaba, some of them accessible only by canoe, who complained of being blocked from the forest's resources by fences and by a special squadron of Green Police (*Força Verde* in Portuguese) who patrol the park to ensure no trees are taken. While the intention is to provide protection against poachers, in reality the police end up harassing and in many instances arresting residents for using, say, a tree to build a roof on their shack or to build a canoe—two real-life instances I encountered while traversing Guaraqueçaba's latticework of waterways and remote villages.

Several arrested villagers ended up finding an unlikely defender—Antonio Machado, a retired bank lawyer who is now the mayor of Antonina, the nearest town to the reserve, who has defended, pro bono, several of the arrested villagers. Located on a bluff overlooking the Bay of Paranagua, Antonina is a simple fishing village—buoyed slightly by the occasional wandering tourist, but a mostly sleepy place of colorfully painted stucco, a church, and a budding collection of artists, there for the cheap rents and incredible scenery. The town, he said, has been inundated with refugees from the reserve, who can no longer sustain themselves under the new restrictions. "Through those conservation projects," he said, "they created a poverty belt around our town."

The three companies that invested in the carbon rights in Guaraqueçaba did so anticipating that the United States would someday establish a cap-and-trade program of its own, and they could use the forest as an offset to their emissions. Those hopes were dashed when the US Congress voted down that effort in 2009 and the companies were left to stash the offsets or sell them on the voluntary market.

The Guaraqueçaba reserve was one of the first experiments in protecting forests with polluters' money. A lot was learned. And the challenges it raised, alas, continue to bedevil the efforts of those who would like to find a way to trade carbon footprints in one place for forest preservation in another.

The tensions about preserving the world's rain forests as carbon sinks have focused on REDD. In the regulated market the European Union established for reducing greenhouse gases, it offers polluters the chance to obtain offset credits for planting new trees on forest-land that's been cleared or degraded, but it excludes the use of credits based on preserving trees that are already there. The EU excludes the use of avoided deforestation offsets for three key reasons. First, it is difficult to produce reliable measurements predicting future increases of CO_2 sequestration in trees thanks to a wide number of variables, including species, elevation, and changing conditions on the ground. Second, monitoring the forest is technologically challenging; it's difficult and expensive to prevent the "leakage" of deforestation from protected to unprotected areas. Third, as we have seen, the practice has sparked widespread opposition from indigenous groups of forest dwellers, like those in Guaraqueçaba and elsewhere.

Instead, there's been a boom in forest carbon offsets in the voluntary markets, which operate independent of any government emission targets. In 2011 and 2012, private US, European, Japanese, and Australian companies paid out more than $450 million to purchase the carbon rights to tropical forests, offering the prospect of sucking up the pollution they create back home.[7]

But those have been voluntary offsets, not subject to government oversight affirming their veracity. There is, though, one exception—the state of California.

Needless to say, not all deforestation takes place in far-off lands, as I was reminded by Christopher Potter and his satellite views. "Let's not just beat up on everybody else," Potter said after our tour of the Borneo forests. "Let's have a look at our own house." He adjusted the satellite's cyclops eye across the United States. There was the light green of the Adirondacks; and then over the Midwest, mostly yellow, a legacy of the drought that took a serious toll on American photosynthesis when it devastated the corn and soybean crops that year. Finally, we landed on the West Coast, Northern California, to the forested lands around Mendocino. There, you had ribbons of brown circumnavigating the green of the redwood forest, a sign of encroaching logging trucks.

In 2013, California became the first, and thus far only, government in the world to formally approve the use of avoided deforestation offsets as part of a carbon trading program. Potter had steered us to a bird's-eye view of the forest that contains the first approved offset in the state—nineteen thousand acres of redwoods around the town of Willits, three hours north of San Francisco. Funds from California polluters like PG&E would enable the owner of this land to resist pleas from timber companies to turn those trees into logs, and instead sustainably manage the forest and protect wildlife habitat and watersheds. That's the theory. The world will be watching how the California approach actually works. As the first such offset to be government-certified, it will serve as a test case of whether reliable measurements of carbon sequestration can be made over time, and whether logging that would have happened there will simply move somewhere else. Critically, there are few people actually living inside that section of the Northern California redwood forest—unlike many other areas of the world where forests are people's homes. It will certainly be a lot simpler to administer than the forests in Chiapas, Mexico, and Acre, Brazil, where people do live and sustain themselves and which the state hopes to add to the list of offsets in the coming years.

Of course, deforestation is not happening in a vacuum. We consume many of the products that come from deforested forests—a concept that has become so widespread that it is now known as embodied deforestation. Citizens of the EU and the United States together account for more than 50 percent of the consumption of global crop and livestock production based on deforested land. Nor are we the only ones: Rising contributors to consumption are Russia, responsible for 15 percent of forest destruction associated with beef production in the Amazon, and China, which accounts for more than 40 percent of Brazilian Amazon deforestation associated with soybean exports.[8] And it's a good bet that all of us, in one form or another, are consuming the palm oil that comes in multiple forms (processed and packaged food, fuel, industrial products) and has led to replacement of virgin rain forests by plantations across millions of acres in Indonesia and Malaysia, according to the Rainforest Action Network. In essence, we the buyers are contributing to the deforestation that the climate negotiators, satellite operators, anti-logging police, government officials, and the rest of us are trying to stop—a metaphor, perhaps, for the fossil fuel cycle we're trying to break.

Back in the pink zone on Werner Goncalves's computer screen, on the day following the bust of the jungle kilns, our convoy of police was on the road again, following the lines from his GPS. Once again we hit the dirt roads and remote highways, passing the alternating rhythms of the threatened forest: Lush, bursting foliage gave way to the grasses that are signs of land recovering from deforestation, followed by newly tragic tableaux of stumps and dirt.

Then we heard a rumble in the distance. We sped up. Heading our way came a convoy of four logging trucks, in a place they were not supposed to be. The police set up a blockade on the road, and the trucks downshifted to a stop. One after another, the agents climbed aboard the trucks—each of them loaded with a dozen or so logs, beautiful reddish hardwoods. The drivers were cited and told to drive immediately to the nearest police station. Our convoy of logging trucks and state police set off for an hour's drive to a brick outpost just off the two-lane highway. There, in a parking lot where two other flatbed trucks sat already emptied of their contraband cargo, the agents began what was clearly a familiar routine: One climbed aboard the logs, calling out the figures to another, who noted them in a legal citation. The drivers, and presumably their bosses, would be subject to a fine depending on the total volume of the seized logs, which would likely amount to thousands of reales—several thousand dollars—per truckload. That would be nothing compared with the potential profits to be made. How much was one of them worth? At least a thousand dollars on the black market, an agent told me.

So is one thousand dollars the price to keep a single tree standing? Or is it just one dollar, as for tree #129 in Guaraqueçaba? Multiplied by the infinity of tropical forest trees on the planet, the United Nations Framework Convention on Climate Change (UNFCCC), facing this broad spectrum of estimates, said that it would cost ten to fifteen billion dollars a year for the next ten years to significantly reduce deforestation rates around the world. The finances are daunting. Over the five years since the Copenhagen commitments, about a tenth of that ideal number, some five billion dollars, has been channeled into programs aimed at discouraging deforestation and recovery from forest degradation since 2010, most of those funds coming from the governments of the EU, United States, Germany, Japan, the UK, and Norway.[9] The Norwegians have, by far, been the steadiest contributor, channeling more than a quarter of that total from their North Sea oil profits into fighting deforestation in the Amazon and slowing the transformation of the Southeast Asian rain forests into palm oil plantations.

Nor is the clock stopping: The official UN and EU policy presumes that the rate of deforestation needs to be cut in half by 2020 and totally stopped by 2030 in order to slow the rate of the warming climate to 450 parts per million of CO_2.

Nor is the direct destruction of trees by humankind the only threat they face. Climate change itself is altering the ecosystems of tropical forests. The IPCC suggests that the combination of increasing temperatures and decreasing rainfall could have a slow-motion impact on soil moisture levels of tropical forests, impacting species diversity. There have been outbreaks of drought never seen before on the northern edges of the Amazon and in other areas of the earth's tropical belt.

So forests offer a perfect portrait of the perfect storm triggered by climate change. They are in the middle of the pincer: Eroded by the very processes already under way for which they offer some kind of cushion, and destroyed by untethered economic forces that do not equate the value of the forest to its trees. Putting trees into a transaction with fossil-fuel-intensive companies highlights this fundamental dynamic—for what those companies do, what they sell, is undermining the conditions they are hoping to preserve.

Which brings us to the ultimate undervalued product—one of the most significant contributors to climate change, the substance that powers the automobiles that companies like General Motors and Chevron are trying to trade for all those trees: oil.

CHAPTER 5

Carbon in the Tank

Oil

On a bookshelf above my desk, I have a rock that's about the size of a fist. It's ridged and bumpy, and covered in oil. I've been carrying it with me for about a decade, after collecting it from a remote beach in Galicia on the northwest coast of Spain.

The beach, about fifty miles north of the border with Portugal, is barely more than a rocky outcropping in the Atlantic, an isolated place with a sharp cold wind even in the spring. Six months before I slipped the rock into my overcoat pocket, on the night of November 22, 2002, a ferocious storm off that coast had tossed the *Prestige*, an eight-hundred-foot-long single-hull oil tanker, like a toy boat. As she lurched in the violent waters, a wave smashed into the right forward hull and the three-foot-thick steel blew open—"like a sardine can," a rescue worker later recalled. After the captain's SOS, the Spanish Coast Guard sent a helicopter to pick up the nineteen crew members. Then the *Prestige* sank about thirty miles offshore. Out from the hull came viscous cascades of oil: Seventy-nine million gallons of crude washed onto a thousand miles of coast, all the way up to the beaches of southwest France. Satellite photos taken by the French research agency CIDRE show the oil spreading from the *Prestige* like spindly black veins in the circulatory system of the Atlantic.

The *Prestige* unleashed one of the worst environmental disasters in history—at least until the *Deepwater Horizon*, BP's oil derrick, exploded in the Gulf of Mexico in April 2010. When the oil started gurgling out of BP's underwater pipes, after the explosion killed eleven people who'd been working on the rig, a sense memory of Galicia returned.

I didn't go to the Louisiana coast to watch, but I did, like many of us, watch in horror from afar. It rapidly became clear that oil spills are not very different. In fact, they are interchangeable. The *Prestige* was carrying a refined version of what the *Deepwater Horizon* was pumping from under the ocean: same substance, different location. By way of contrast in the fossil-fuel-catastrophe sweepstakes, the *Prestige* spill was far bigger than the *Exxon Valdez* crack-up in 1989, which spilled some fifteen to thirty million gallons of oil (the exact number is still in dispute) along the coast of southeast Alaska, less than half of the *Prestige's* toxic load. The *Deepwater Horizon* blew them both away—unleashing over its excruciatingly public effusion some 210 million gallons of oil. I was standing on the Atlantic coast of Spain, but I might as well have been standing on the coast of the Gulf of Mexico, south of New Orleans, or, for that matter, of Prince Edward Sound in Alaska.

I went to Galicia to investigate the causes of the *Prestige* accident for the PBS newsmagazine show *FRONTLINE/World*, and spent many hours clambering over those oil-splattered beaches.[1] Even in the wind, the air on that isolated beach had the sickening smell of a gasoline station. All I could see in either direction—to the south, toward Africa and to the north, toward Ireland—were thousands of rocks, millions of them, stretching into infinity, covered in the black gooey crud that had been carried in the cargo holds of the *Prestige*. An assortment of volunteers from as far away as Italy and the Czech Republic were still scraping rocks and boulders with spoons in a quixotic effort to remove the vile substance.

The closest town was Corcubión, about ten miles away, which for hundreds of years has been sustained by the abundant fish of the

North Atlantic. The town is renowned as the source of a marine delicacy unique to this area called percebes, a barnacle that clings to the side of the rocks and is a prized feature of Galician cuisine. The waves that crashed upon those rocks were tainted with oil: There would be no percebes that season, nor any other fish caught along the Galician coast. The marine environment was severely damaged, and tens of thousands of fishermen suddenly lost their source of livelihood all along the coast. Tourists who normally flood this area in the spring and summer stayed away.

The economy of Corcubión was devastated. I met the town's mayor, Rafael Mouzo, who with his trim beard and bushy mustache looked like a character out of Cervantes. As we walked along the stone balustrade overlooking the town's port, lined with idle fishing boats stranded by the blacklist on Galician fish, Mouzo could barely contain his rage at the disaster that had befallen his town. He told me that the oil spill was like an act of war. "This," he exclaimed, "was an act of terror, a criminal act! We need an international tribunal to judge . . . all those responsible for the spill." Currents carried the *Prestige*'s vile cargo northward, and in fishing communities along its path there was a similar sense of having been besieged by an evil, destructive force that seemed to come from nowhere and let loose its demons upon their shores.

The cleanup costs quickly mounted into billions of dollars as the Spanish government sent out hazmat crews in insulated white suits to power-spray the beaches with high-pressure hoses. Some of the earth's most pristine beaches were plunged into a scene out of science fiction.

I've held on to that rock for all these years so I would not forget the image and feel of the grotesque environmental and economic destruction wrought by that sunken tanker, and the price that is paid for our reliance on a fuel that at every turn wreaks collateral damage while it powers our economic might. That Galician rock, perhaps the only physical legacy remaining of the sunken *Prestige* here in the United States, tells a story of the acute effects of oil unleashed upon the sea and upon the land. Who could have guessed we'd see the same infinity of oil-splattered rocks along a thousand miles of the Gulf Coast? Well, we all could have guessed. Oil spills have not been rare events. And it's a good guess we'll see more of them, somewhere, because oil is the Esperanto of the world's fossil fuels, speaking the same language everywhere.

Half of all global oil use is for transport, and most of that for automobiles, according to the International Energy Agency. There's only one real distinction between the oil once in the hold of the *Prestige* and the oil I use every day. After a successful journey across the sea, "my" oil was refined into gasoline that provides the energy for the steady beat of the pistons in my car. It pumps away through my fourteen-year-old Saab's plumbing, and leaves its CO_2 fumes behind. So that's the other reason I held on to that rock: It reminds me that I, like all of us, am tied to the oil carried by the *Prestige*—except my shipment of oil and its associated greenhouse gases ended its journey in my gas tank and not on the beaches of Spain. The CO_2 embedded in the *Prestige*'s oil from all those millions of years of decay at the center of the earth went straight into the atmosphere without detouring through my car. All that BP oil went straight from the center of the earth onto the beaches of Louisiana without detouring through a refinery and onto an oil tanker. And the unique circumstances of the BP spill, in the middle of the ocean, unleashed huge quantities of methane, a greenhouse gas twenty-five times more potent than CO_2, from under the ocean floor—"the most vigorous methane eruption in modern human history," a Texas A&M University scientist told the press.

The effects of climate change, of course, are occurring far more slowly than the overnight destruction that follows an oil spill. But where the CO_2 comes from, or what it powered in the intervening sequence of connections into our gas tanks, is an incidental detail in oil's journey. The rock above my desk is coated with the substance that we rarely actually see as it travels up wells in distant locales like Saudi Arabia or Venezuela or one hundred miles offshore from

Louisiana, into pipes and tankers and finally into the refineries of America that process it into the energy that fuels our cars.

We, too, are contributing every day to what amounts to an ongoing oil spill into the atmosphere of our planet.

I pull my car into a Chevron station near my house in Berkeley. The prices at my local station are not much different from anywhere else in town. The people who run the station are friendly, stuck behind their Plexiglas booth, immigrants from Ethiopia. They're charging $4.15 a gallon. Wow, that bites, as it does every time we go to the gas station.

Alas, that's another fake price in the economic hall of mirrors—in which the price reflects not actual costs but the reflected glare of the margins necessary for oil companies to sustain hefty profits. Perhaps this is the harshest news in the struggle to shift our habits away from fossil fuels: Gasoline should be a lot more expensive than the "expensive" gasoline we've been paying for. Nor are the true costs invisible; they're hiding in plain sight. You just have to do the math.

According to the EPA, transportation is the second most significant contributor, after industrial sources, to greenhouse gases in the United States. More than a quarter of the nation's greenhouse gas load comes from moving ourselves and our goods around—all those trips down the block or across the country add up. I could have filled my gas tank about ten million times with the amount of oil spilled in the BP disaster. (I don't drive that much, so you'd have been using some of that gas, too.)

Personal vehicle use is responsible for almost half of those transportation emissions: 787 million tons of CO_2 in 2010, to be exact, according to the EPA. The average American, said the EPA, uses 557 gallons of gas a year driving around; and each of those gallons emits twenty-five pounds of CO_2. Twenty-five pounds, as was pointed out by Sarah Terry-Cobo, my former researcher and a journalist affiliated with the Center for Investigative Reporting, is about the weight of a single cocker spaniel. That's 557 cocker spaniels' worth of CO_2 a year, from each of us with a driver's license.

Over the course of reporting this book, I traversed the state of California by car, driving several times into the Central Valley and as far south as Los Angeles—at least one thousand miles of driving on that 2000 Saab. I emitted twenty-five thousand pounds of CO_2. It would take about a tenth of an acre of forest to absorb that carbon footprint over the course of a year if I were to offset it in one of those Brazilian forests. Terry-Cobo's animated film *The Price of Gas*[2] walks us through the various stages of how a single gallon of gasoline's greenhouse gases journey into the atmosphere, using figures backed up by studies from the EPA and various academic institutions across the country. The pumping of oil from Saudi Arabia and transport by tanker to the United States, she shows us, results in two pounds of greenhouse gases per gallon. The refining, at a Chevron refinery in Richmond, about fifteen miles from where I write: another 3.5 pounds. And then there's the pumping into my gas tank, which includes evaporation at the pump, and driving: That's another nineteen pounds. Thus, 80 percent of the total emissions per gallon comes from us, the drivers.

My car gets about twenty miles to the gallon, so my Saab consumed fifty gallons of gas during my travels across the state. That comes to about two hundred dollars for gas. I should have paid a lot more, according to the calculations of those who are beginning to apply a price for gasoline that approximates its actual cost.

Of the top twenty-five industrialized countries, the United States has the lowest price for gasoline. In Germany, for example, in 2013 the average price of gasoline was $8.50 per gallon; in Britain, $8.04; in Norway, $10. That's why the price of gas for all those reporting trips should have been about double what I actually paid for it—say, more like about four hundred dollars. And it's not because the gas is cheaper to obtain in the United States. Britain and Norway, for example, have a huge supply of oil just offshore, in the North Sea.

We all get oil mostly from the same constellation of suppliers—the dozen member countries of OPEC, as well as Russia, Britain, Norway, and the United States, distributed to us, for the most part, by the top ten global oil companies.

The difference is that a tax has been slapped onto gasoline in those countries to help compensate for its immense ecological and economic costs. The same goes for airplane fuel: The quantity of greenhouse gases emitted by US airlines is not even counted by the Federal Aviation Administration, and, as we explored in chapter 1, the United States is resisting European efforts to bring prices more in line with true costs. Meanwhile, many EU countries—notably Germany, Britain, France, and the Netherlands—have been channeling the extra funds generated through gasoline taxes to begin the process of lowering the greenhouse gas emissions of European cars, devising less greenhouse-gas-intensive fuels, and mitigating the effects of climate change.

The Organisation for Economic Co-operation and Development (OECD) has concluded that environmental taxes account for approximately 2.4 percent of GDP in Europe, compared with 0.8 percent in the United States.[3] Jason Scorse, an associate professor of environmental economics at the Monterey Institute of International Studies, has calculated that the 1.6 percent difference between the two figures—given the roughly comparable sizes of the European and American economies (Europe's is actually slightly larger)—would provide the US government with $240 billion annually in extra revenue. Even a portion of those funds, he said, "could be applied to the monumental costs of climate change and associated environmental problems associated with gasoline." Consider that figure in the context of the two devastating and costly events of 2012 alone whose intensity has been linked to climate change: the fifty billion dollars in public money apportioned by Congress to the East Coast for recovery from Hurricane Sandy; and the eleven billion dollars in damages, apportioned through the federal crop insurance system, from the midwestern drought. Even a minimal addition to the already minimal tax on US gasoline would have gone some way toward covering those costs.

The federal gasoline tax in the United States now stands at 18.4 cents per gallon—a figure that has not changed since 1993, when President Clinton, egged on by his vice president, Al Gore, jammed it through a reluctant Congress. The price of gas was roughly $1.75 to $2.00 in the mid-1990s—so what was a roughly 9 percent tax then is now half that as the price of gas doubled.

A team of engineers at Carnegie Mellon University and Arizona State University concluded that conventional cars over their life span come with an average of two thousand dollars in greenhouse-gas-related costs that are "paid by the overall population rather than only by those releasing the emissions and consuming the oil."[4] And those costs don't come just from climate-induced weather events. They affect our health care system, too. The American Public Health Association, for example, estimates that the health costs from respiratory diseases and premature death due to automobile-induced air pollution amount to eighty billion dollars a year. Even these numbers don't tell the whole story. Beyond storms and health are the more slow-moving, but potentially far more expensive, long-term impacts from greenhouse-gas-induced climate change. A gas tax, say costing (at the top end) roughly two hundred dollars a year over the decade-long life span of your average car—and assuming the EPA's average American gas consumption of 557 gallons per year—would begin to compensate for those costs. Instead of each of us paying those costs up front, depending on how much we use our cars, we all now pay collectively, irregardless of whether we even drive a car.

Indeed, gasoline prices have risen significantly over the past ten years, but all those extra billions did not go into the development of less greenhouse-gas-intensive fuel technologies. Rather, the price increases were channeled into the profit margins of the major oil companies. The years of 2011 and 2012 taught us a critical economic fact: When the price of gas goes up, consumption may decline incrementally, but oil company profits do not drop. In fact they did the opposite: Over the course of 2011, as gasoline prices started their most recent rise, the country's top five oil companies (Chevron, ExxonMobil, Shell, BP, and ConocoPhillips) earned a combined profit of $137 billion, despite *reducing* their levels of production. In 2012, as gas prices continued to rise and stayed above four dollars, the top five oil companies hit another one-hundred-billion-plus year—while the actual quantity of oil they delivered to the market declined by at least 5 percent. Nor, concluded a report by the minority Democratic Party staff of the House Committee on Natural Resources, did the companies use those funds to hire more Americans; in fact, the top five oil companies shed at least four thousand employees while reaping those record profits.[5]

And there's another twist to this perverse economics: The US oil and gas industry receive billions of dollars in taxpayer subsidies—a kind of expensive phantom limb left from the days early in the last century when oil companies had to actually be encouraged to seek out ever more exotic locales for their oil prospecting.

Parsing the precise quantity of subsidies to the fossil fuel industries can be difficult because the calculation involves a thicket of gifts large and small, from tax breaks to outright R&D support to hidden price props like the Strategic Petroleum Reserve. Estimates of the amounts vary widely, so let's start with the most direct and conservative, which is quite eye-opening: The OECD offers us an Excel sheet documenting at least $3.4 billion in subsidies to the oil industry in 2011, and another $3 billion to the gas industry.[6] The programs include everything from tax breaks and favorable depreciation treatment for crude oil exploration to a sales tax exemption for the purchase of new oil and gas equipment. (This is the low end of the estimates and excludes some two billion dollars to assist low-income earners to meet their electricity and heating bills, as well as the lost income from below-market rates for drilling on federal lands and in the Gulf of Mexico.) In 2011, the Democratic staff at the House Committee on Natural Resources estimated that the oil and gas industries were slated to receive $43.6 billion in subsidies over the following decade, including things like the foreign income tax credit, which lets oil companies off the hook for a portion of profits generated in lower-tax-rate jurisdictions. Between 1918 and 2009, according to an assessment by DBL Investors, a venture capital firm based in San Francisco, the oil and gas industries received an average of $4.8 billion in government subsidies annually.[7] Whichever number you choose, the taxpayer-funded subsidies buttress the oil companies' bottom line every year, and enable the world's primary greenhouse gas pollutant to be sold more cheaply than its actual cost—adding to the perceived price advantage of fossil fuels over renewables. Although the oil industry has been on a steadily profitable trajectory for decades now, congressmen and presidential candidates who have collectively received more than $365 million in contributions from the oil and gas industry since 1990[8] have been loath to repeal the subsidies (President Obama tried several times, but was rebuffed, with a few small exceptions, by the Republican-controlled House of Representatives).

Globally, the London-based Overseas Development Institute (ODI) estimated that the top eleven "rich country emitters" (a designation that has come to be known as the E11, including the United States, Germany, Britain, Canada, Japan, Australia, and Russia) collectively spent $74 billion in taxpayer funds on fossil fuel subsidies in 2011. That comes down to seven dollars for each ton of greenhouse gases emitted.[9] Even the International Monetary Fund has entered into the lambasting of fossil fuel subsidies, which it sees as not only an inducement for greenhouse gas emissions but also a violation of free-market principles, creating an unjustifiably favored status for fossil fuels. Removing global fossil fuel subsidies, the IMF said, would reduce CO_2 emissions worldwide by 13 percent, and produce "major gains" in economic growth by freeing the funds to be used more productively in the economy.[10]

So, not only are we paying for the oil companies' overhanging costs—covering the climate consequences of their most significant waste product, CO_2—but we're also paying to facilitate their genera-tion of those costs: Refining, drilling and extracting oil and gas account for at least 390 million tons of greenhouse gases every year in the United States, the second biggest source of emissions after coal-fired power plants.[11]

Thus, the matter of subsidies raises a key question: If we're paying already to support the oil and gas industry, why not channel those funds instead toward alternatives that will lead to the development of more resilient, renewable, and less destructive ways to power our engines? A hundred years ago, arguably, government subsidies were needed to help promote the rapid development of a revolution-ary new fuel source, at a time when few understood its long-term environmental consequences. But now we understand. Today taxpayer subsidies serve to inflate the profit margins of the fossil fuel companies, and in fact encourage the extraction of ever more difficult-to-reach sources of oil—drilling deep beneath the ocean floor, for example, and fracking, which involves breaking apart the earth to extrude oil from the rock. So in effect we pay twice: once to inflate oil companies' profit margins, and again for the damages those fuels cause to the planet through their emissions of green-house gases.

Thomas Schaller, a political science professor at the University of Maryland, suggested that we consider how the flow of capital would differ if we instituted, say, an extra twenty-cent-per-gallon tax on gasoline. The benefits, he argued, would be multiple: We could see a decrease in greenhouse gas emissions due to decreased gas consumption; channel billions of dollars into the government's ability to respond to the rising costs triggered by climate change and other environmental consequences of oil; and provide resources to further develop innovative technologies to wean us off our reliance on petroleum. Rather than feeding the bottom line of ExxonMobil and the other titans of oil, we could channel our funds to adapt to and mitigate the effects of the greenhouse gases that the oil compa-nies—with our complicity—have been sending into the atmosphere.

An increased tax on gasoline would clearly have a disproportionate impact on people with lower incomes. But there are ways to ensure that they are not the hardest hit. Congress has repeatedly entertained, and defeated, several proposed bills that would impose a gradual increase in the gasoline tax coupled with tax rebates or other forms of compensation to ensure that low- and moderate-income people are not shouldered with the burden. Such an initiative could ensure that the excess price paid for gasoline would be channeled into programs from which we all benefit, rather than the money we pay in other taxes being channeled to enhance the bottom line of oil companies.

Increasing attention to the greenhouse gas implications of petroleum-based fuels is creating new divisions in the once solid alliance between oil companies and car manufacturers, which have for decades together been the primary obstacle to tightened environmental standards. In 2012, the car companies agreed, after years of protests, to President Obama's proposal to raise fuel efficiency standards for American cars to an average of 54.5 miles per gallon by 2025—and have been pouring millions of dollars into developing new engines for electric and biofuel-powered vehicles. When California instituted a low-carbon fuel standard, which requires that the "carbon intensity" of fuels used in the state be reduced by 10 percent by 2020—a goal accomplished largely through the addition of biofuels to the gasoline mix—the measure was immediately challenged in court by the oil companies and some new allies, biofuel producers in the Midwest concerned that their corn- and soybean-based fuels were rated as higher greenhouse gas contributors than the grass- and waste-based biofuels more prominent in California. They were not joined in the challenge, however, by the auto companies. General Motors, for example, which underwent a resurgence after the Obama administration kicked in fifty billion dollars to prevent the company's collapse, directed its engineers at research-and-development centers to design engines that could meet the requirement.

"We're on the opposite side of the oil companies in the battle over the low-carbon fuel standard," said Shad Balch, a former analyst on climate policy for former governor Arnold Schwarzenegger, and now head of future product policy and communications for GM. The company's long-term interest, he said, lies with reducing the dependence of the next generation of cars on the erratic supplies, and volatile prices, of oil. "We want to see a new standard so we can take advantage of it," he told me. "The first company with a no-gas car wins!" California is GM's biggest market, as it is for all the auto companies, so the cars that meet the state's relatively tough standards will ultimately be sold throughout the country. Other car companies, like American Honda, have set a goal to reduce greenhouse gas emissions from cars by 32 percent from 2000 levels by 2020. Part of that is by increasing production of e-vehicles.

The sales of electric cars, reliant purely on the electrical grid for power (as opposed to hybrids like the Prius, fueled by a combination of electricity and oil), have skyrocketed as battery technology improves and the price comes down: In 2010, 345 individual e-vehicles were sold in the United States; by 2013, that number had leapt to over 50,000.[12] That's still a small number relative to cars on the road, but all the major US and European car companies, in addition to new entrants like Tesla Motors, are banking on increasing numbers. The auto industry is global, so what happens elsewhere can also feed into the move toward less-oil-reliant cars. In Europe, for example, where consumers daily face those steep gas taxes, car manufacturers also face an EU Fuel Quality Directive that requires reducing greenhouse gases from transport by 60 percent (from 2009 levels) by 2018, creating an incentive that is helping lead to rapid growth in electric vehicles there. And European cars are already about a third more fuel-efficient than their American counterparts[13]—which has the effect of not only decreasing the car-by-car release of greenhouse gas emissions, but also decreasing expenditures for gasoline. Then there's the world's fastest-growing car market, China, which in 2013 announced a national goal of five million electric cars on the road by 2020, a development that could have a transformative impact on the industry.

Critically, however, automobile manufacturers have little interest in how green the grid is that their cars rely on. What do plug-ins plug into? For a car company, it doesn't necessarily matter how the electricity is generated, as long as it's there. "Building a market for electric vehicles has nothing to do with greening the grid," commented Robert Langford, head of the electric vehicle division at Honda, whom I interviewed during an e-car promotional event held at PG&E headquarters in San Francisco.

But for electric cars to truly have significant impact, they must ultimately draw power from a grid that relies on less destructive sources of energy. What is the source of the electricity? How green is the grid? It turns out to vary widely depending on your location.

In California, electric vehicles emit about five times less greenhouse gases over the course of their life cycle than gasoline-powered ones. That's because California's Renewables Portfolio Standard—the strongest in the nation—requires that utilities obtain a third of their energy from renewable sources by 2020; the state already draws a significant portion of its power from wind, solar, nuclear, and hydro. In states such as Colorado and Wyoming, where the reliance on coal is greater, that ratio drops to somewhere between two and three to one, according to a joint study by the Natural Resources Defense Council and the Electric Policy Research Institute.[14] Twenty-seven states have some version of a Renewable Energy Portfolio Standard, so the greenhouse gas impact of electric vehicles will vary depending on comparative greening-of-the-grid state by state. The Department of Energy's budget for 2014 included $615 million to increase the use of solar, wind, geothermal, and hydro sources of energy to further green the grid, and $500 million dollars in research in "cutting edge vehicle technologies" designed to wean American drivers off oil. Meanwhile, there's been a boom in states like California, New York, Connecticut, and others in what's known as distributed energy, referring to energy created through wind, solar, and other renewable sources that's fed into the grid—a phenomenon spreading across the country that presents a direct threat to the long dominion of monopoly utilities over who gets to generate, and profit from, electricity.

However sporadically the greening of the grid is taking shape, one thing is clear about cars powered by electricity or by biofuels, hydrogen, or other technologies on the horizon: They don't use oil. Just as distributed generation threatens the lock of utilities on our electricity system, electric cars present a direct threat to the lock that oil companies have had on how we move.

One of the great ironies of the struggle to assign the proper price to oil is that the industry itself acknowledges the risks and consequences of its key product. "Rising greenhouse gas emissions pose risks to society and ecosystems that could be significant," ExxonMobil tells its potential investors in a filing with the US Securities and Exchange Commission. Chevron, which was the biggest funder of an unsuccessful statewide vote to repeal California's landmark greenhouse gas law, informs visitors to its website that the use of fossil fuels is "a contributor to greenhouse gases," and that such gases are leading to climate change, "with adverse effects on the environment." ConocoPhillips expresses an even higher level of concern: "ConocoPhillips recognizes that human activity, including the burning of fossil fuels, is contributing to increased concentrations of greenhouse gases in the atmosphere that can lead to adverse changes in global climate." Shell goes so far as to advocate that "CO_2 emissions must be reduced to avoid serious climate change." As for BP, the company's website offers some remarkable candor, as if speaking about another industry entirely: "More aggressive, but still plausible, energy policy and technology deployment could lead to slower growth in CO_2 emissions than expected, with greenhouse gas emissions from energy use falling after 2020—but probably not enough to limit warming to no more than two degrees centigrade."

There is clearly something bothering the oil companies about climate change—a backbeat of uncertainty, what the insurance industries call pure old-fashioned risk: the literal risks from rising seas that could inundate coastal refining facilities, for example, and more to the point the risk that the government will awaken to the immense costs of climate change and slap a significant price on carbon. It helps explain why in December 2013, ExxonMobil and four other oil companies indicated that they were starting to integrate a cost of from forty dollars a ton for carbon (Shell and BP) to sixty dollars a ton (ExxonMobil) in their in-house accounting. As of early 2014, there was no place on earth where the price even approached the low end of those numbers, but it illustrates how far ahead the companies are in anticipating a growing public demand to price fossil fuels more accurately. Theoretically, if CO_2 takes on an extra sixty-dollars-per-ton price, that could make investments by the companies in alternative energy sources more attractive by comparison, conceivably even steering them toward wind, solar, and other renewable energy sources. Absent a mandated price, though, it's business as usual.

There may be another purpose, as well, to the oil companies' expressions of awareness. Now that we know that they know, then it's up to us whether we want to continue consuming their product. The revealing passages on company websites may be a kind of equivalent of the cancer warnings on cigarette packages: It's far more difficult to sue a tobacco company on the ground that they kept the damages from cigarettes secret. The same may hold for oil companies: If we know what our purchases lead to—mainly, more loading of the atmosphere with CO_2—then what's their responsibility?

In fact, for oil companies, the question of what to do, of being part of the process of slowing down the accumulation of CO_2, strikes at the essence of their raison d'être—which is why they have so aggressively attempted to block any serious climate legislation in the United States. As the Carbon Tracker Initiative pointed out in its seminal report, *Unburnable Carbon*, the oil companies deep investment in the status quo can be explained by the trillions of dollars' worth of assets they possess, in the form of drilling permits and concessions, just awaiting the time when they can be pumped and processed and delivered into the holding tanks of a ship like the *Prestige* and finally into the gas tanks of our cars and turned into money. This is the specter that looms over the world, for in order to increase our odds of keeping the global temperature rising no more than two degrees Celsius by 2050, the IPCC calculations require that 60 to 80 percent of oil company assets remain exactly where they are, in the ground, unburnable. Without an enforceable price on carbon, that is unlikely to happen.

———

During my reporting about the sunken *Prestige*, I was able to determine why the tanker sank and why no one was held accountable for the series of shortsighted decisions that led it to sea and the massive amounts of oil it spilled onto the Spanish coast.

I spoke with the man who captained the ship immediately before the captain who had command that fateful night when the ship went down in an Atlantic storm off the coast of Galicia. Efraizos Kostazos had been working as a tanker captain for a quarter century, he told me one afternoon in the Greek port city of Piraeus, and had never before seen a ship in such bad shape as the *Prestige*. He had command of the *Prestige* as it sat for six months in the Russian port of St. Petersburg as a lightering vessel—a stationary gas station for outbound ships. The American Bureau of Shipping in Houston, Texas, had recently conducted an inspection and signed off on the vessel's seaworthiness. But Kostazos was concerned about the tanker ever being taken out to sea again. He sent a fax to the owners back in Athens, as well as to the ship's inspectors. In his urgent communiqué, Kostazos expressed serious reservations about the ship's condition and identified a litany of serious structural flaws in the *Prestige*'s aging steel hulls. He recommended they be repaired before she was taken on another voyage. Rather than fixing them, however, the owners fired Kostazos—and sent another captain, Apostolos Mangouras, to replace him. The *Prestige*'s owners were a Greek family, the Coulouthroses, who owned six ships at the time.

At twenty-six years old, the *Prestige* was one year past the age that even the International Association of Tanker Owners considers safe. But there were no laws to prevent the aging tanker from being taken out to sea again. The Coulouthros family's maritime business in Piraeus sent word to Mangouras that a shipment of oil awaited in the Lithuanian port of Vilnius. The tanker's days as an immobile ship were over. Mangouras steered the *Prestige* out of St. Petersburg, picked up its cargo of Russian oil in Vilnius, and headed toward a buyer in Singapore. The tanker passed through the Baltic and North Seas and into the Atlantic Ocean, where it hit the storm off Galicia. The waves blew holes in the areas of the right forward hull that had been identified as vulnerable by Efraizos Kostazos.

The cleanup of the beaches and compensation to fishermen cost at least five billion dollars, paid for primarily by the European Union and Spain. The owners' insurance policy contributed twenty-five million dollars, not even 1 percent of the total. It turned out that the trapdoors of the maritime system ensured that there would be no further responsibility for the damages.

There was no international tribunal, as called for by Rafael Mouzo, the mayor of Corcubión. In fact, the saga of the *Prestige* was one of dodging responsibility. When the Spanish government filed suit in the United States against ABS to reclaim the damages, the inspection company argued that responsibility lay first with the flag state, the Bahamas. Stuck in a catch-22 of maritime law, a federal judge in New York agreed and dismissed the case. The Bahamas has minimal resources to inspect the hundreds of ships flying under its flag and no legal authority to impose its will on international shippers had it even been inclined to do so; it claimed, as ABS did, that final responsibility lay with the owners. The owners had ensured they would have no financial liability because they registered the *Prestige* as a "one-ship company" with the maritime authority in Liberia—the legal home of more than half the world's oil tankers. The company's sole property, the *Prestige* itself, was at the bottom of the sea—and thus no longer had any value. There were no assets to be claimed. It was as if the entire maritime system conspired to ensure that the *Prestige* had never existed.

But the *Prestige* had existed—as the residents of Corcubión and the entire Galician coast know very well. The buck was passed—in this case, onto the citizens of Spain and Europe, who paid for the monumental damages caused by that one ship on that one night in November. Though the oil itself was Russian, it rapidly became the problem of Galicia, of Spain, of Europe.

Now let's spin that scenario one more time, this time in the Gulf of Mexico. The parallels seemed eerie as the major players began tossing around blame for the *Deepwater Horizon* like a live grenade. BP blamed Halliburton, which built it, and Halliburton blamed BP, which after all was operating the drilling platform and gaining the profits from its dredging of the sea. Ultimate legal responsibility, according to the principles of maritime law, lay with the flag state with which the *Deepwater Horizon* was registered; being a maritime vessel, the *Deepwater Horizon*, like the *Prestige*, had to carry a flag indicating its technical "home." The flag the *Deepwater Horizon* was flying was that of the Marshall Islands—a former territory of the United States that has even fewer resources to enforce maritime standards than the Bahamas. They contract out to private companies to conduct inspections. And the inspector of the *Deepwater Horizon*, paid for by BP, was the American Bureau of Shipping, the same company that had provided the clean bill of health to the *Prestige*. Four months before the *Deepwater Horizon* exploded, ABS certified that the drilling rig was in safe working condition.[15] Even though the drilling occurred in American waters, the US Coast Guard has no authority to review the safety of what's technically known as a Mobile Offshore Drilling Unit, or MODU, registered in any country other than the United States. As the Government Accountability Office stated in a report to Congress, US "inspectors do not have authority to review a self propelled, foreign flagged MODU vessel security plan."[16]

To complete the circle, the *Deepwater Horizon* explosion and leak would happen in ways that had been partly predicted by an internal report produced by the owner of the rig, Transocean (BP rented it from them for five hundred thousand dollars a day). Just weeks before the accident, inspectors for Transocean concluded that the port side of the rig was "in bad condition and severely corroded . . . worn and in need of being replaced." BP used it anyway.

This time, it was as if the entire maritime system was conspiring to ensure that neither BP nor anyone else would have to assume liability for the environmental damages caused by its oil drilling in the Gulf of Mexico. Under pressure from the Obama administration, BP agreed to provide a twenty-billion-dollar fund to compensate businesses and residents for lost income and homes along the most directly impacted areas of the Gulf Coast. But the total estimated costs, as of 2013, are approaching fifty billion dollars—costs that are the subject of numerous lawsuits, and a substantial amount of which we taxpayers will end up paying. And the residents of the Gulf Coast are still living with the consequences.

The experience of the massive and expensive environmental catastrophes that followed the oil spills on the Spanish coast and on the Gulf Coast seems to encapsulate the entire question of Big Oil's contribution to the costs of climate change: They contribute mightily to the planet's load of greenhouse gases, yet, thanks to the trapdoors like those that have shielded the owners of the *Prestige* and the operators of the *Deepwater Horizon*, are not held responsible for its consequences.

The spills from the *Prestige* and the *Deepwater Horizon* just made the oil visible. On most days, when we are, thankfully, spared such overt tragedies, the greenhouse gases from oil just rise into the atmosphere quietly. To the atmosphere, the location of spilled oil off the Galician coast or in the Gulf of Mexico is incidental. From up there, where the CO_2 collects and wreaks its damage, it's just another day down here on earth, with the bustle of cars and trucks, drills and tankers, and another fill-up for my Saab. Another day, another oil spill.

CHAPTER 6

A Tale of Three Cities
The City

T
he Allegheny River pulses through the city of Pittsburgh, Pennsylvania, like a muscle. Along its banks, John D. Rockefeller punched the nation's first oil well. Along its waters traveled the steel that became the backbone to our railroads and skyscrapers, and the coal that fired up the factories fueling America's twentieth-century industrial might.

Then in the 1980s and early 1990s, the steel started leaving. Today some two decades after the flight of the last mill, the Allegheny, no longer the transport system for American industry, has been transformed into a different kind of symbol—one of the modern "green" city. The river has been cleansed of many of the most toxic substances that formerly poured from those factories; now its meandering flow is featured in municipal brochures against a glittering downtown skyline that hosts one of the highest concentrations of green buildings in the United States.

I followed the Allegheny one winter afternoon along a tree-lined route, now known as the Greenwalk, that once carried workers to the factories. From the riverbank, I could only imagine the clamor and the heat and the smoke that had inspired a 1950s-era columnist to describe Pittsburgh as "hell with the lid off."

Just blocks off the Greenwalk the hulks of those steel mills are still visible. They, too, have been transformed—into condominiums with a river view, gourmet restaurants, music clubs, and boutiques. Hipsters hang in the cafés that were once steelworker meeting halls. Young couples pass by with their kids in tow; bicyclists whiz past on designated lanes. Where there were once generations of families reliant on forging heavy metals, the city's new residents are firing up innovations in its burgeoning high-tech and biomed industries, backed by an assortment of world-class universities.

All those "creatives" are certainly generating plenty of intellectual capital; they have sent Pittsburgh on a greening binge. From the Greenwalk to downtown, Pittsburgh has been attacking its polluting gases as if its survival depends on it—which it does. The city and surrounding communities once produced more than half of the steel in the United States. After the industry bottomed out, a coalition of businessmen, city planners, and environmental engineers staked out a development plan that positioned Pittsburgh as a hub of innovation in ecologically oriented design.

Downtown, the skyscraper windows are angled to maximize natural light, heat is piped in from thermal pools deep underground, and solar panels line the roofs far above the bustling sidewalks. The city has a full-time sustainability director charged with shifting its energy sources away from fossil fuels by means of mass transit, urban planning, and municipal procurement policies that place a value on low-carbon alternatives. The city's water treatment system—channeling water of varying quality, from rain to sewage, into multiple uses—is considered a model even for other eco-conscious cities like San Francisco. Major property developers agreed to halve their 2003 carbon footprints by 2050; the city now has the highest concentration of LEED-certified buildings in the country.

Even the US Steelworkers, born in Pittsburgh more than a century ago as one of the country's first industrial unions, now has a huge banner draped from its downtown headquarters reading GREEN

JOBS—which it has been promoting in the hope of landing new employment for its members. The city's transformation has been so complete that the G-20—representing developed countries, all of which have experienced similar declines in manufacturing—held its yearly conference here in 2009 and highlighted the city's green strategy as a model for the post-industrial way forward. Pittsburgh, once home to the industrial empires of Andrew Carnegie and Andrew Mellon, is now one of the greenest of midsized cities in America, according to *The Economist's* Green City Index, and even rates as one of the "most livable" places in the country—a designation that would have been unthinkable a decade ago.

"There was a saying here in the old days," commented Court Gould, the executive director of Sustainable Pittsburgh, a coalition of planners, environmentalists, and business leaders that helped with the blueprints for the city's transformation. "A father would take his son out to their yard, look back toward the mills and the smoke rising over them, and tell him, 'Look there, that's my job. That smoke there, that's money.'"

The outlines of those mills and warehouses, now perked up into lofts and assorted high-tech workshops, loom in the distance from his office. Gould, a lifelong resident, continued: "Not anymore. Pollution no longer has the smell of money. Now it's the smell of costs. It's the smell of someone not paying attention to the bottom line. It's a sign of inefficiency, of, 'What's wrong with the management?'"

Pittsburgh's greenhouse gas emissions plunged from the days when all those factories filled the sky with waste gases that would, according to accounts of the time, turn entire afternoons into twilight. "Back then in the '50s and into the '80s, no one was thinking about climate change, and no one was asking about emissions of carbon dioxide," said Aurora Sharrard, who served as the first president of the city's Climate Initiative, a collaboration among the business, municipal, and scientific communities to devise emission reduction strategies. From 1900 to 1970, while Pittsburgh and surrounding Allegheny County assumed their place among the centers of American industry, the area experienced a steady annual increase in its industrial greenhouse gas emissions, from fourteen million tons to thirty million tons. Then, according to a team of researchers at Carnegie Mellon University, emissions started to decline. As industries fled the area, so went their greenhouse gases: Industrial emissions declined by half throughout the area between 1970 and 2000;[1] and they've continued on their downward trajectory. By 2008, the city of Pittsburgh, the financial, cultural, and political center of the region, was down to about 6.8 million tons.[2]

By 2013, the city was well on the way toward its goal of reducing emissions 20 percent from 2005 levels, and aims for progressively steeper declines over the decades to follow. Public transit has been expanded, subsidies for solar and thermal energy have promoted an expansion of small- and large-scale renewable energy for residents and businesses, and waste disposal services have been improved to enhance recycling and other energy-saving measures.[3] In a country that is reliant on local improvisation, with few legal guidelines from the federal government, Pittsburgh is considered among the leading urban climate innovators.

Pittsburgh lost its manufacturing base, and it's a far nicer place to live as a result. It retooled its efficiencies, cast off the harmful by-products of manufacturing, and refashioned itself as a city far more reliant upon brains than on brawn.

So whatever happened to all those greenhouse gases that once came spewing from Pittsburgh?

They did not disappear.

Welcome to Guangzhou, a city of ten million on China's southeast coast. The freighters that come into the port here, and in the surrounding Guangdong province, are loaded with a container about every second—some forty million crates a year, all carrying Made in China goods to the United States, Europe, and around the world. Industrial clusters throughout the province are home to more than a thousand steel manufacturing and trading companies. They produce skyscraper girders, auto parts, appliances, ships, refrigerators, and even American bridges—steel products that once were made in Pittsburgh and other midwestern cities during America's century of industrial dominance.

Guangdong is also, in the UN's estimation, one of the top ten carbon-emitting provinces in a country that is itself the leading emitter. Some eight thousand miles from Pittsburgh, the CO_2 that used to come from that city now fumes into the atmosphere from Guangzhou. It wasn't just Pittsburgh's manufacturing jobs that migrated to China; the greenhouse gases associated with them went, too. And so did the jobs and CO_2 from those myriad factories in other US cities that have either outsourced their production or been crushed by Chinese competition. Between 1990 and 2008, report a team of researchers led by the Center for International Climate and Environmental Research in Oslo, the emissions embodied in products imported by developed nations from developing ones—which for the most part means China—grew by as much as 17 percent every year.[4]

The World Bank estimates that about a quarter of all the production in Guangdong province, and indeed throughout China, is destined for export to the United States, Europe, and Japan. The resulting greenhouse gases are, in economists' terms, known as embedded emissions. They are the pollution backstory to the goods we consume.

All told, according to a study by Chinese scientists published in the journal *Atmospheric Chemistry and Physics*,[5] the residents of this churning industrial center of China have a per capita annual footprint of 7.8 tons—quite a bit more than the average Pittsburgher. But there's a story, too, hidden in those numbers. Only 6 percent of Pittsburgh's emissions, according to that city's Climate Inventory, come from industrial sources. Yet emissions of the Chinese industrial sector account for 56 percent of the total—almost ten times higher, as a percent of the total, than those of Pittsburgh.

The huge discrepancy between the industrial emissions of Pittsburgh and the industrial emissions of Guangzhou, which started trading places as centers for steel production in the 1980s, suggests that the lifestyle choices of Pittsburghers have not changed as much as the economic support system, based on greenhouse-gas-intensive manufacturing, changed all around them.

What the Chinese numbers tell us is that legions of urban residents, some of them people who have literally replaced those steelworkers, have a far smaller personal footprint as a percent of the overall total than do their Pittsburgh counterparts. As Chinese consumption grows that footprint will grow, but at this stage and for some time to come it is production, and not consumption, that accounts for the overwhelming bulk of Chinese emissions. Indeed, as much as 60 percent of China's exports are manufactured by China-based affiliates of multinational corporations, many of them American and European.[6] The Chinese, in short, are producing greenhouse gases on our behalf.

So the greenhouse gases that were once American are now Chinese. China was responsible for about 10 percent of the world's greenhouse gas emissions in 1990; by 2013, their global contribution was more like 30 percent. But greenhouse gases end up in the same atmosphere wherever they are generated. Does it matter where they come from? On one level, it does not. But on another more fundamental level—the level on which we decide who is responsible for reducing those emissions—it does, significantly, matter. Climate scientists say that to avoid a potentially catastrophic rise of four degrees Celsius by 2050, each of us should ideally be emitting no more than two tons of greenhouse gases annually. Consider this finding from the Carnegie Institution for Science at Stanford University: Americans' per-capita footprint would jump by 2.4 tons annually if their consumption—mostly of goods made in China—is taken into account.[7] Pittsburgh is about average on the scale of American consumption, so all those green Pittsburghers would have to add those tons to their carbon load.

Globalization has flipped the calculus on the central question of who is accountable for greenhouse gas emissions. Richard Feldon, a San Francisco–based urban planner, headed up a team of scientists working with the International Council for Local Environmental Initiatives to devise a set of 2012 guidelines for American cities seeking to limit their emissions. He told me that including consumption in their calculations was one of the most controversial issues they faced over the three years it took to identify the primary sources of greenhouse gas pollution in American cities. That's because it blurs the line between our contribution as consumers and industry's contribution as producers, and gives a new way to understand our "greenest" of green cities.

"Let's say Pittsburgh still had its industrial base, and that steel from Pittsburgh was being used in a city like San Francisco," Feldon explained. "Well, it would be unfair to say that San Francisco, under that scenario, is a greener city than Pittsburgh." The same equation, he said, applies to Pittsburgh and Guangzhou—or, say, the United States and Europe, jointly the world's biggest consumers, and China, the world's biggest producer. It also means that when you do the numbers, the United States switches position with China to become the world's top emitter of greenhouse gases.

Most of us are urban dwellers—70 percent of the world will be by 2020—so reducing those emissions, city by city, is one of the fundamental challenges the world faces in devising a new energy system that keeps greenhouse gases to at least livable levels. Just because emissions aren't happening in our backyard doesn't mean that they're not ours—a reality that at least one of the world's cities is facing head-on.

Welcome to Manchester, birthplace of the industrial revolution. This city in the northwest of England was at the center of Britain's industrial rise, and has experienced a trajectory similar to Pittsburgh's ever since. Take a walk down Oxford Road, as I did one rainy spring afternoon, and you'll pass red-brick buildings with terra-cotta embellishments that were the headquarters of massive textile combines driving what was once the world's mightiest industrial economy. Now the street is lined with high-profile global brands and boisterous cultural centers; like its American counterpart, Manchester has one of the highest concentrations of solar panels and green office buildings in Britain. The weather is about as gray and misty as Pittsburgh's, too. And like Pittsburgh, Manchester has dropped from being one of its nation's leading greenhouse gas emitters to a center for high-tech innovation, and is host to a cluster of leading universities conducting cutting-edge research into renewable energy. The legacies of both cities are interwoven deeply into the evolution of greenhouse gases and their contribution to climate change.

Pittsburgh hosted the world's first commercial oil well and was the center for American steel; Manchester was home to the world's first steam-driven factory, which of course was reliant on coal.[8] In the eighteenth century that coal-fired energy was put into the service of processing the vast amounts of cotton that Britain was obtaining from its colonies in Asia and North Africa. Cotton was a key ingredient in Britain's imperial rise, and Manchester was the key to turning all that cotton into textiles. By 1850, Manchester was widely considered a model for the modern industrial city. Alexis de Tocqueville went there for a visit shortly before his legendary foray to the United States, and reported on the city's combination of ingenuity, industrial invention, and vile living conditions. "From this foul drain," he wrote, "the greatest form of human industry flows out to fertilize the whole world. From this filthy sewer pure gold flows."[9] Friedrich Engels, the economist who partnered with Karl Marx, wrote his seminal book, *The Condition of the Working Class in England*, here while overseeing a cotton mill owned partly by his father.

Textiles were to Manchester what steel was to Pittsburgh—a ticket to technological innovation, and the huge quantity of greenhouse gas emissions that came along with it. Indeed, one can see

Manchester as having assumed the industrial greenhouse gas contribution on behalf of the British colonies, which were expected only to send raw materials to the mother country for processing and manufacture. The legacy of Manchester's onetime dominance of the world textile trade can be seen in unexpected ways, such as the exotic patterns that the mills incorporated into bland British designs to sell as processed clothing back to the colonies. That included the paisleys that slipped into Western fashion in the 1960s as Manchester designers learned to sell to the consumers of India.

Like other imperial cycles, Manchester's dominion came home to roost. Today the textile companies are long gone—many of them returned to India and China, where they now make the patterns to satisfy Britons' stylistic demands. By the 1990s the city had rocketing unemployment as its industries were vacuumed to Asia. Practically an entire generation of workers in Manchester were compelled to leave the city or live on the dole. Their greenhouse gases went with them. Call it a post-colonial update to the era of climate change.

Then in 1996, in the center of the downtown, the IRA set off the most powerful urban bomb explosion in its history of conflict with the British government. Some 110 people were injured and the neighborhood around the city hall decimated. And it was then, I was told by Sarah Davies, who is in charge of environment strategies for the consortium of ten local governments known as the Greater Manchester Combined Authority, that the city was compelled to decide how it wanted to rebuild itself.

There was a "shift in the mind-set," she said, among the city's traumatized leaders—who met as the world prepared for negotiations over what would become the Kyoto Protocol. Manchester would return to its role as a center for technological innovation, but this time that innovation would be adapted to the emerging vision of the new low-carbon economy. Like Pittsburgh, it staked its future on brains, not brawn. It would shake its reputation as a polluted industrial center past its prime, and instead redesign its economic core, Davies said, "to be resilient to the global changes." One key element of those changes was the growing awareness that the climate talks would, soon enough, bring about a world in which carbon had a price. "There's the sense," she told me, "that we created the energy-hungry economy. And now we have some responsibility for finding our way out of it."

───────

Pittsburgh and Manchester's industrial history may be similar, but the way that the two cities deal with their greenhouse gas accounting is not. The divergence can be seen in the two cities' official climate strategies. According to Pittsburgh's Climate Inventory, "Emissions resulting from many personal and business-related activities and decisions that might be evaluated in an individual, carbon-footprint-style inventory are excluded from a city-level GHG inventory approach." In other words, the city is not counting the carbon embodied in the goods and services its residents consume, or generated by their travel.

By contrast, the long-term plan published by the Manchester City Council calls for accounting for, and reducing, emissions by city residents "wherever those emissions take place."[10] These embodied emissions include the energy needed in the growing and transport of food; the extraction and processing of oil used by the city's automobiles and factories; the emissions generated through the manufacture of electrical devices and appliances; and estimates of aviation emissions by residents flying out of town. Adding these consumption-based emissions adds roughly 30 percent to each citizen's carbon contribution to the atmosphere—for a total of some 15.7 tons per person, according to a 2012 estimate.

The Greater Manchester Combined Authority, representing some three million people in the city and surrounding communities, launched an initiative to reduce the city's footprint not only at home, but also in the countries producing the goods consumed by its residents. Its room to maneuver is of course limited, but procurement policies now favor imported goods with lower greenhouse gas impacts than their competitors, and the city has embarked on an effort to educate employers and homeowners on precisely why purchasing goods closer to home, and reducing energy usage, is good for the city's economy, as well as for the planet.

Davies, whose office sponsors what's known as the Carbon Literacy Project, identified a central principle of that effort—providing a motive beyond some vague sense of well-being, "like we were all coming from some experiment in the '70s," or that 'saving carbon' is somehow akin to well-meaning impulses like saving money for a rainy day. "What is so intrinsic to energy, to carbon, that makes people want to save it? You have to ask, 'What is it that people want?' They want to earn more, pay less, have a decent quality of life, that's what people aspire to. So carbon literacy must be put through these channels. They need to see 'prosperity' as 'green.' Something that makes their lives better. You have to tie it to the aspirational goals of people or you will have no effect." Manchester is trying to make saving carbon desirable. The city's long-term aim is to reduce emissions 41 percent from 2005 levels—a goal that Davies said makes economic as well as environmental sense.

"Our competitors are other British or European cities," she added. "Having this target makes us more attractive to investors." Siemens has established a green technology center in Manchester, as have numerous other Japanese and European firms; and textile companies are being lured back to the city from India and China, attracted by new fuel-efficient ink and dye technologies, developed with the assistance of city subsidies to the university, as part of the burgeoning move toward sustainable textiles. This has a double bonus: creating jobs in the city and seriously shortening greenhouse-gas-intensive transport costs. Between 2007 and 2012, the city created thirty-seven thousand new jobs in its evolving green economy, representing £5.4 billion ($7.5 billion) in money passing through Manchester that would otherwise have gone elsewhere—an "elsewhere" that likely would have been using far less energy-efficient technology than is now the norm in that city. The economy of Greater Manchester grew 4 percent in 2012, fed largely by the infusion of green investments, Davies said, at a time when growth in the UK was flat.

In this way, consumption-based greenhouse gas accounting prompts a fundamental twist to our common understanding of globalization: The world may be flat when it comes to production, but when it comes to greenhouse gases it is definitely round—and we are at the other end of the circle. "You cannot decouple production from consumption," commented Cindy Isenhour, an associate professor of environmental studies at Centre College in Lexington, Kentucky, who has conducted extensive research on embedded emissions for the Stockholm Environment Institute.

Of course, national governments are limited in the influence they have over the production practices of other countries. Consumption accounting presents a troubling challenge for governments accustomed to acting within the traditional confines of national jurisdiction, even for those at the forefront of efforts to fight climate change. In June 2012, trouble came when the question of responsibility for those outsourced greenhouse gases caused an uproar in the British House of Commons.

In that month, Prime Minister David Cameron announced that Britain had decreased its emissions by 20 percent from 1990 levels, and was well on its way toward meeting its target of 80 percent reductions by 2050. He was greeted by a barrage of criticism from scientists and municipal officials who'd done the research on embedded emissions or thoroughly digested its implications. One of Cameron's own scientific advisers, David Mackay, an Oxford University physicist, said the government was laboring under an "illusion," and that in fact Britons' consumption of greenhouse-gas-intensive products had increased by some 18 percent at least during the time period cited by the prime minister—as shown in life-cycle analyses of the rising consumption of imports. That claim just about nullified the government's claims of progress.

The government's own Department for Environment, Food & Rural Affairs concluded that while Britain's territorial carbon footprint had indeed shrunk 12 percent between 1993 and 2004 (a different but roughly equivalent time frame to that cited by the prime minister), the emissions related to imports had risen during that same time period by 59 percent.[11] Most of those emissions were linked to imports from China. The House of Commons Energy and Climate Change Committee held a series of contentious hearings on the findings in the fall of 2012, at which the chairman of the Greater Manchester Combined Authority and a host of scientists from the University of Manchester testified. The committee concluded that not only were the government's figures misleading but that "the UK's consumption is driving up territorial emissions in other countries." They published a study suggesting that consumption accounting would reduce China's emissions tally by as much as 45 percent. A representative of the Department of Energy and Climate Change, which has primary responsibility for the UK's climate policies, was asked how he could explain the discrepancy between the rise in consumption and the prime minister's claim of falling emissions. The official responded, defensively, in writing: "The increase in consumption-based emissions is associated with globalization and consumption patterns, not caused by domestic environment policy." The government demurred that it could have little influence over the pollution caused by the manufacture of products abroad that the British population consumes.[12]

"The government thought that all those reductions had come about because they were more efficient," recalled Alice Bows, a researcher at the Sustainable Consumption Institute at the University of Manchester, whom I spoke with several months after she'd testified at the Commons hearing. "It looked like they were ahead of the game. Then this research suggested there are other components to the picture. Our findings made them uncomfortable."

Their findings should make all of us uncomfortable. This enormous blind spot in greenhouse gas accounting is shared around the world. A survey by the Stockholm Environment Institute concluded that virtually every developed country, including the United States, had underestimated its emissions by 20 to 30 percent by failing to include increased consumption.

"We are talking about changing habits and lifestyles," said Bows. "That messes with you. People think they have a choice. They're attached to the idea that they have a choice of how to live. But choice is always constrained, limited by circumstances, financial and other realities. So we need to alter the concept, the circle of possibilities included in 'choice.'"

And so this saga of three cities brings us back to what is certainly the most discomfiting question of all the discomforts that climate change brings. Follow all those many circuits of production, follow the trail of greenhouse gases rising into the atmosphere, and you will ultimately land upon each of us, making our choices about what we consume and from where. At what stage we act—at the point of production or the point of consumption—is a matter of how directly we face our contribution to climate change, now intertwined so deeply already with our economic reality.

The city of Pittsburgh and the city of Manchester are both signatories to a commitment signed by more than three hundred cities around the world that have committed to seriously reducing their greenhouse gas contribution. Both are largely unheralded leaders among the world's cities in attempting to face the challenges of climate change. Manchester, however, is at the forefront of attempting to account accurately for its outsourced emissions—and suggests a forward-looking approach to what is bound to become an ever more volatile question as individual jurisdictions—cities, states, nations, regions—attempt to wrestle with their emissions in the absence of an international accord. Such tensions are already in play even within the United States, where California has been far ahead of the rest of the country in contending with its greenhouse gases. That led to a portentous legal struggle, in which California echoed the role of Manchester—which in turn, of course, echoes the role of the European Union's effort to regulate transnational airplane emissions.

The state of California's low-carbon-fuel standard requires that the average greenhouse gas intensity of transportation fuels be reduced by 10 percent from 2008 levels by 2020. The state determines a fuel's greenhouse-gas-emitting potential through a life-cycle analysis of every stage of the fuel-production process; for instance, in the case of ethanol, from the emission consequences of clearing land for growing feedstocks like corn to the refining process and on to the final stage of combustion in vehicle engines. The state was challenged by a coalition of out-of-state oil, ethanol, agricultural, and trucking companies, which claimed that the law violated principles of interstate commerce by subjecting out-of-state energy companies to California law. A federal district court ruled in the companies' favor, asserting that the state could not apply its own standards of carbon intensity across state borders. The state appealed, though, and in September 2013, the Court of Appeals for the Ninth Circuit—as high as you can go before the Supreme Court—reversed that decision and affirmed that California has the right to apply its own life-cycle criteria to fuel sold in the state.

This is not just a matter of legalese. For the first time, a court affirmed that extraterritorial principles within the United States could be applied because greenhouse gases have been found by the EPA "to endanger the public health and to endanger the public welfare of current and future generations." The danger presented by greenhouse gases trumped the usual prohibitions against differing standards from state to state. In the words of the Ninth Circuit Court's decision: "One ton of carbon dioxide emitted when fuel is produced in Iowa or Brazil harms Californians as much as one emitted when fuel is consumed in Sacramento."[13] Danny Cullenward, a lawyer and environmental resources scholar at the UC Berkeley Energy and Climate Institute who contributed to an amicus curiae brief supporting the state, saw the immediate national and international implications of the court's decision. "The case," he said, "planted a flag in the ground. The question was, 'Are you responsible for the smokestack or for the consumption?' Because we're in a global economy, you have emissions coming from multiple different points, from places that are not necessarily correlated with the location of the product being consumed."

This question of extraterritorial jurisdiction will continue to be one of the basic ways in which climate change, a built-in transgressor of frontiers, challenges our old ways of understanding borders of authority—whether on the state, national, or international level. It's certainly at the heart, also, of the aviation case we explored in chapter 1. In that case, though, the United States and Europe continue in their battle because there is no international body other than the International Civil Aviation Organization with teeth to enforce authority over transnational emissions. Such are the challenges when regulating greenhouse gases in a world of improvised carbon prices. And as long as this improvisational vein continues, such conflicts will continue to surface. The EU, for example, is considering the prospect of implementing a carbon tariff on goods that are not subject to the same emission restrictions as European goods—a move that would undoubtedly end up as a major battle in the World Trade Organization.

Whether in Pittsburgh, Manchester, or somewhere else, residents of prosperous cities and nations are most likely leaving their carbon trails over other cities or nations, often somewhere far away, and often less prosperous, like Guangzhou. Those trails are a network of ghost emissions—displaced footprints plodding in the no-man's land of the atmosphere, blurring the evidence as the questions over who has responsibility for them becomes ever more intense.

So in a sea of CO_2 displacement, how do we pursue meaningful reductions? That's the question that, as we'll see, reverberates through the surreal world navigated daily by a new breed of financial traders on the climate crisis's very own Wall Street, the carbon markets.

CHAPTER 7

The Clean Dark Spread

Pricing Carbon

*F*rom London, arguably, the whole climate conundrum commences. From here came the eighteenth-century principles of enclosure—the parceling of the public commons into private properties that lay the foundation for replacing farms with factories. From here came the philosophers, like Adam Smith, who articulated a rationale for the free market and the industrial revolution that it served. And from here, in the city's boisterous coffeehouses, came the invention of modern financial vehicles that sustained the rapid expansion of that revolution.

This boom box of a city was where the processing of decayed fossils into fuel began its journey into the center of our economic order. Manchester had the textile factories; London had the ideas, the political will, and the money.

Given its history as midwife to the fossil fuel economy that set us on this path of atmospheric chaos, it stands to reason, then, that London is now the financial center of the effort to reduce its most destructive side effects. One of the world's most ambitious economic experiments is centered here—the effort to implant a price on carbon through the mechanisms of the free market. Though it's not so free, and it hasn't worked so well as a market—nor was the hub originally anticipated to be here—the carbon market reflects a largely unheralded accomplishment of the annual climate negotiations since the Kyoto talks of 1997. Every new set of negotiations, as fitful and sporadic as they have been, has brought us a slow and steady introduction to the idea that greenhouse gases come with a cost that must be paid.

From London, also, comes the economist Nicholas Stern, who was the first to put hard figures on the costs of climate change. A school of economists have followed in his footsteps, who have articulated principles of environmental accounting that look not only at the up–down trajectory of company profits, but at the broader and largely unaccounted-for costs to the health of the planet that lie behind those profits. One of the pioneers in this effort is Trucost, a London-based environmental accounting firm that works for British businesses, the government, and the United Nations. Trucost's work lies behind several seminal studies by the House of Commons Energy and Climate Change Committee, as well as the efforts by companies like Virgin and Puma to reduce their carbon footprints.

Alastair MacGregor, the company's chief operating officer, commented that current accounting practices, developed from the mid-1800s to the mid-1900s, are rapidly being outpaced by the scale of environmental crises. "They were not developed to deal with what we've seen over the past couple of decades, with concepts of environmental risk and climate change." In 2013, Trucost completed a study estimating that the excess costs from lost environmental services due to damage to the earth's natural capital from climate change amounts to some $2.7 trillion (roughly the size of the United States' 2013 national budget). Many of the world's largest emitters would no longer be deemed profitable if their actual environmental costs were counted.[1] This radical way of understanding corporate profits—laying on top of them the grid of externalized environmental costs—lies at the root of the Kyoto accord and other climate agreements, which can be understood as an effort to start putting into financial form our growing understanding of environmental risk and its consequences.

To do so, though, requires contending with a heretofore unseen challenge. As opposed to regulating, say, smog, with pollution emanating from a limited number of sources and locations and impacting a specific number of usually urban people, the primary contributions to climate change come from an infinite variety of sources, and the impacts are experienced by literally everyone. A single coal refinery in, say, Charleston, West Virginia, sends greenhouse gases into the atmosphere that add to the gases already there from similar refineries in, say, South Africa, China, or England. And those gases go on to affect every one of us, though at vastly different levels and dimensions.

By 2030, the United States, for example, could lose as much as 2 to 3 percent of GDP dealing with climate-related events and impacts if current emission levels continue[2] (a level roughly equivalent to the Defense Department's share of the GDP), while in neighboring Mexico, the impact could be more like 6 percent,[3] and significantly higher in heavily populated Mexico City, according to an assessment by the National Center for Atmospheric Research in that city. That's not only because of where Mexico is located—in latitudes of greater weather volatility—but also due to the relative resilience of physical infrastructure. That disproportionate dynamic is reflected globally: The impact in North Africa is more heightened than in Europe, not only because of its latitudinal location but also because its population of largely subsistence farmers has few resources to draw upon to recover, which also heightens the chances that residents will flee, putting stresses on the places where they seek refuge—such as Europe or, in the case of Mexico, the United States.

Apply that in multiple variations across the planet and you start getting a picture of the complexities of trying to pinpoint cause, effect, and responsibility. The causes—greenhouse gas emissions in one place—have effects that can be down the block or thousands of miles away, or both at the same time. Those effects can unfold years apart or in the same week. There is also the cruel truth that many of the countries that feel the impacts most profoundly—for example,

Bangladesh, with a huge population arrayed across lands that are at sea level, or the Pacific island nations that are faced with inundation—have contributed only minimally to the problem.

The world has been attempting to funnel these many diverse implications and causes of climate change into one fundamental economic equation, to start paying the costs of externalities now rather than waiting for the catastrophes to happen, when it is far more expensive. "Externalities have been the biggest free lunch in the history of the world," said Pavan Sukhdev, who worked for seventeen years as a senior banker at Deutsche Bank and now consults with Trucost, the United Nations, and other clients on helping to identify hidden environmental risks and costs.

It was to begin accounting for those externalities that London's carbon market experiment was created. Through this new financial vehicle, polluters would start paying for all those "free" lunches. Major greenhouse gas emitters were identified; emission targets were established; the world would have one price for carbon, and businesses could compete equally upon this newly elevated playing field. That, anyway, was the hope.

Patrick Birley, a tall and energetic South African, greeted me in the Bishopsgate neighborhood of central London some five years after he'd launched the world's biggest carbon trading floor, the European Climate Exchange, or ECX. Birley exudes the high-octane optimism of a man who has spent his career speculating on the prospects of commodities. Carbon is only the latest. A native South African, Birley was educated at the London School of Economics. After Nelson Mandela led his country out of global isolation and back into the global economy in the 1990s, Birley returned home to create the South African Futures Exchange. He returned to London in 2005 to launch the ECX; by 2007 it was open for business, and he was on the cutting edge once again.

To Birley, carbon is as tangible as any other commodity. "They're like pork bellies!" he exclaimed. "Except these are emission credits." We were looking out on his trading floor, where digital renditions of Europe's greenhouse gases flickered across the screens. Indeed, there is certainly the same potential for profit in betting correctly on the price volatility of pigs and carbon.

But as much as this market resembles pork bellies, there is a critical difference between the two. The world's five greenhouse gases—known collectively as carbon—have been turned into a financial commodity on a market that is supposed to last until it eliminates the need for itself. Somewhere in the slaughterhouses of the world a pork belly—market shorthand for the tasty underside of a pig, the source of bacon and pork chops—actually exists. With carbon, however, there is no delivery of a tangible anything. Rather, it's an investment in a unit of one ton of greenhouse gases that is not emitted into the atmosphere. And whereas the more money is channeled into pork bellies, the more pigs will come to market, with carbon the dynamic is quite the opposite: The more money put into carbon, the less of it there's supposed to be. As the system took root and the economy grew, the amount of money channeled into carbon rocketed from sixty-five billion dollars in 2007 to more than two hundred billion dollars' worth of transactions by 2013.

Unwinding how that rapid growth happened offers a lesson in creating a market from scratch: Invent a demand, then supply it. The story is actually not so different from those ingenious merchants who at the cusp of the industrial revolution in the London coffeehouses invented new funding devices, like new forms of risk-sharing insurance, to finance themselves and their commercial enterprises. Except this time the objective is to give the world's primary pollutant, the unanticipated (at the time) side effect of those inventions, a price that begins to reflect its actual cost.

The Kyoto Protocol established a goal of reducing emissions from 1990 levels by 20 percent by 2020, and 80 percent by 2050—and mandated creation of the market to get there. Only developed countries that ratified the protocol participate, including the twenty-eight members of the European Union plus Norway, Iceland, Liechtenstein, Australia, New Zealand, Canada, and Japan (though Australia and Canada are in various stages of retreat from the requirements of the protocol). The procedures are roughly the same for all protocol signatories, but the EU Emissions Trading System—the EU ETS—is by far the largest.

The market's main currencies are the EU Allowance (EUA) and a parallel trade in UN-certified offsets, Certified Emission Reductions (CERs), which can be used to meet up to half of a company's emission obligations (more on those in the next chapter). Each of these confections represents a permission to pollute one ton of greenhouse gases.

Creating a commodity out of the air—literally—is a complicated enterprise. It starts with more than eleven thousand of Europe's most greenhouse-gas-intensive industries—utilities, refineries, steel, chemical, and cement manufacturers—representing 45 percent of Europe's greenhouse gases. Each is issued an emission limit, based on an average of the top ten most energy-efficient emitters in each industry in the years 2007 and 2008. The companies then are issued emission allowances based on that number. Emissions above their limit must be purchased at the going rate for carbon on markets like the ECX. The Europeans added airlines to the market mix as well in 2012; so it is in these markets, too, that air carriers buy allowances for the emissions coming from all those planes.

April 1 is a big day for Europe's greenhouse-gas-intensive industries: On that day every year, each one of those companies must submit its emission totals to the EU's Directorate-General for Climate Action and submit an allowance for each ton. Any company that emits more than the benchmark level, the "cap," must have purchased more allowances, or face a fine of one hundred euros (about $135) per ton of excess emissions.

The cap gets progressively tighter as we move closer to 2020, and thus the price of allowances is supposed to grow progressively higher. As of 2014, the cap is scheduled to drop by 1.7 percent every year, a number that is designed to hit a 20 percent emission reduction target from 1990 levels by 2020. Non-emitters—in other words, financial speculators—are also encouraged to enter into the market and wager on the carbon price in order to maintain liquidity, thought to be key for leveraging the price upward, providing an incentive for companies to shift away from fossil fuels.

From this convoluted route grows the supply (emission allowances) and demand (emission targets) that greases the circuits of the carbon market. At the outset, the new invention captured the imagination of key constituencies. The financial community and many, but not all, environmental NGOs saw it as a way to use market forces to channel the profit motive toward increasing the price of carbon. It would also, they thought, create a more equitable playing field for renewable energy sources. And it appealed to industries, which were looking to minimize the costs of compliance with the new regime; buying credits would be cheaper than lump-sum investments in emissions-reducing infrastructure. These differing motives seemed, at first, to be elegantly aligned, though soon enough they would end up on a collision course.

For a time, carbon was the fastest-growing commodity on the planet. Growth was explosive. In 2006, 450 million tons of carbon were traded over the ECX circuits; by 2012, that figure had tripled to 1.3 billion tons and to over 2 billion tons by 2013.[4]

Tons of carbon? As we sat in Birley's office—glass walls, sleek and uncluttered, thoroughly wired—I had trouble visualizing the weight of gases that have wrought such havoc down here on earth. That huge yellow balloon containing ONE TONNE CO_2, floating a couple of stories above Copenhagen during the negotiations in 2009, was about as close as I would ever get to a physical representation of greenhouse gases. Here the gas is not a physical presence. Greenhouse gases are rendered in digits, packaged in one-million-ton bundles of permission to release one million balloons' worth of greenhouse gases into the atmosphere. (Another way to look at that one ton of

CO_2, courtesy of the US Environmental Protection Agency, is as the equivalent of driving 2,160 miles in the average American car, or the annual electricity use of 125 average American homes.[5])

The Pacific Gas & Electric Company, which supplies power to the San Francisco Bay Area and, not incidentally, to my computer, emitted sixteen million balloons' worth of CO_2 in 2013 (16 million tons), according to its submissions to the Air Resources Board, which administers the state's greenhouse gas law. Or take the far larger midwestern utility American Electric Power, the single largest user of coal in the United States; it emitted 122 million of those balloons of CO_2 in 2012. AEP, the company that is attempting to offset its emissions by buying carbon rights to the Guaraqueçaba forest that we explored in chapter 4, is not required to report its emissions to anyone, but offered those numbers voluntarily in its "sustainability report" to potential investors. By contrast, RWE, one of Europe's largest utilities, emitted 198 million balloons, according to its 2012 annual report, in which it detailed the emissions it is obligated to report to the EU's Directorate-General for Climate Action. The German company monitors the minute-by-minute emissions from its dozens of coal and natural-gas-fired power plants at company headquarters in Essen, Germany, and turns to its traders on the ECX to buy the allowances the company needs to continue emitting.

Here on this one question of who is emitting how much, we have a mini portrait of the chaotic and improvised approach to climate pollution. Each of those three companies is among the top polluters in the United States and Europe. And each of them is subject to widely varying regulations regarding its emissions. PG&E reports its California emissions to the state Air Resources Board, which enforces the requirement to reduce emissions in the state to 1990 levels by 2020, a figure that is presumed by state modelers to be a 30 percent cut from what would have occurred without the law. AEP, operating in the Midwest and Southeast, is under no federal or state emission reduction limits. However, the Obama administration did impose standards under the Clean Air Act that require much tighter emission controls on coal-burning power facilities, giving AEP and other utilities an incentive to move away from the dirtiest fossil fuel toward less polluting sources like natural gas. RWE is operating within the world's biggest carbon-regulated market, and must abide by greenhouse gas limits mandated by the European Union to reduce its emissions by 20 percent from 2005 levels by 2020.

I looked out through the translucent panes of Birley's office to the office beyond. The strategy to etch a price for greenhouse gas pollution into the energy supply chain was in full bloom, quiet and entirely virtual. There were no traders jumping and shouting their bids, an artifact of the previous century. Rather, they were focused on the small up-and-down digital tics on their computers representing bets on the price of carbon.

Carbon's fate on the market is tied inextricably to its energy commodity cousins. The lower the price of coal, the higher the price of carbon—because lower coal prices are an incentive to continue using coal-based energy and pay the extra allowance costs, thus increasing the demand, and price, for allowances. The reverse is also true—higher coal prices generally trigger a shift away from coal and thus a decrease in the demand for carbon emission allowances.[6] There is another dynamic at play, too: the price at which energy generated through natural gas—which emits half the CO_2 of coal—becomes more economically attractive than coal. The price for carbon—in essence a penalty for using coal—would have to reach at least forty-five dollars (roughly thirty-five euros) per ton to make that switch economically attractive to power generators.[7] This interplay among coal, gas, and carbon is a continuing theme of the carbon market dynamic. Looked at another way, it also means that the cheaper carbon gets, the less appealing it becomes for polluters to switch to renewables; and the more expensive carbon gets, the more appealing the switch to renewables. There are two other significant factors influencing the carbon price: Extremes of heat and cold, for example, require more energy, and more allowances for utilities. General economic performance also plays a role: When production goes up, the demand for energy follows and the need for allowances increases. The reverse is also true, as the world discovered during the financial crisis that plunged European production between 2009 and 2012.

The limited group of energy traders and policy makers who understand this complex enterprise can make or lose fortunes betting on swings in the price. Henry Derwent, chief executive of the International Emissions Trading Association (IETA), which represents the world's carbon traders and exchanges, explained the term traders use to describe their investment positions on carbon: *clean dark spreads*. That moniker is distinguished from *dark spreads*, used to describe positions taken on fossil fuel commodities. *"Dark,"* he said, smiling wryly, "means it's coal." Add *clean* to the spread and you're betting against the fuel sources that contribute to climate change. The higher the price of carbon, the less attractive the price of coal by comparison with renewable alternatives.

"You need to make the price high enough," said Birley, "so that it hurts. The idea is to make it cheaper to invest in new technologies than to purchase emission credits. You want a price that hurts."

So what is a price that hurts?

Birley said that about thirty-five to forty dollars per ton is necessary for the financial jolt necessary to prompt investment in alternative energy sources. Point Carbon, a widely cited market analysis firm created by the global media giant Thomson Reuters, estimated it would cost more like forty to fifty dollars a ton to induce electricity users to switch from coal to cleaner fuels. Nicholas Stern, in his seminal report for the British government, the "Stern Review on the Economics of Climate Change," asserted that an eighty-five-dollar-per-ton price for carbon would begin to provide the necessary trigger to an energy transformation. Thomas Heller, a Stanford economist and director of the Climate Policy Initiative, told me it should be more like $150 per ton to "truly stimulate innovation."

All those ideal numbers are a far cry from the actual workings of the market. Trading carbon is more like trying to shoot an arrow at a moving target. And the targets are far more the result of the volatile geopolitics of climate change than of fine-honed economics.

The scene for these economic vagaries was set in Kyoto, where in 1997 global negotiators gathered five years after the UN's first alarming report about the economic and environmental consequences of the atmospheric shifts triggered by greenhouse gases. The UN and practically all economists agreed that the costs for slowing the rate of climate change were an order of magnitude less than would be recovering and adapting to it when the symptoms get more severe. Given the uncertainties of science and the risk-based business of economics, however, it was not, and is still not, precisely clear how an investment in reducing greenhouse gases would pay off with a milder impact from climate change in the future. This is best understood in the realms of probability.

Let's make a brief digression into the calculations that underlie how governments think about the principle of paying now to avoid damages in the future. They broach this quandary by applying an economic principle known as the discount rate to our moral obligations to future generations. In the world of climate economics, the discount rate can be calculated by asking a simple question: What are you willing to pay now in order to avoid more costs in the future? The lower the discount rate, the more you are willing to spend to prevent or limit future environmental consequences. Mark Trexler, whose company, The Climatographers, consults with corporations on such principles of risk, puts it this way: "The lower your discount rate, the more you value the future."

This is not just number crunching on the head of a pin. The discount rate feeds into the government requirement (dating back to the Reagan administration) that it perform a cost–benefit analysis before implementing any environmental regulation, including those dealing with climate change. There was no discount rate in the United States specifically applicable to climate change until 2010, when President Obama appointed an interagency task force to come up with one. They reviewed the existing literature on climate risk and devised a discount rate of 3 percent. That rate lies behind the EPA's determination that carbon's social cost amounts to at least thirty-eight dollars a ton, the figure by which it justifies the benefits of greenhouse gas regulations. The figure is in essence an intergenerational cost–benefit analysis, and was a compromise between a much higher rate advocated by industry lobbyists, and a much lower rate advocated by environmental organizations like the Natural Resources Defense Council.[8] (For example, a 5 percent discount rate lowers the social cost of carbon to twenty-seven dollars per ton; a 1.5 percent discount rate raises it to sixty-five dollars a ton). Republicans in the US Congress have wanted no number at all.

Other countries, like the UK and France, use a declining discount rate, starting from a 3 percent baseline—and dropping from there. This means, for example, that the UK foresees the social cost of carbon rising to about one hundred and fifty dollars per ton by 2020, and to as much as two hundred dollars per ton by 2050.[9] Laurie Johnson, chief economist at the Natural Resources Defense Council, argues that the British and French approach offers a far more realistic acknowledgment of the high uncertainties triggered by climate change. "A static discount rate," she said, "does not reflect the kind of uncertainty we're facing. A declining rate is far more accurate in capturing over long time horizons the tipping points and potential for catastrophic risk."

Of course, anticipating future costs by mitigating emissions is a separate issue from adapting and responding to the already significant costs of climate change. We already pay for the consequences of the extreme weather and temperature shifts that have been buffeting the planet. The costs of mitigation require extrapolating into the future—the weighing of odds that are articulated far more precisely today than they were in 1997. Some form of this gamblers' exercise was a backbeat to the calculations of climate negotiators who converged upon Kyoto. They were faced with figuring out a new way to channel the massive amount of resources necessary to avert catastrophic levels of greenhouse gas emissions. How much? Where will the money come from?

Torque is one of those concepts that engineers know is key to creating friction, the friction of change and transformation, which is what the world hoped to trigger in Kyoto. Looming over the talks was the historical record, which shows that CO_2 began being emitted into the atmosphere in quantities that cannot be explained by the earth's normal cycles in the decade between 1780 and 1790—the period corresponding to the beginning of the industrial revolution in Britain, soon to spread to continental Europe and to the newborn United States of America. We in developed countries spent two centuries paying a fraction of fossil fuels' actual costs, thereby hastening our own economic successes. Global demand for fossil fuels is projected to rise by as much as 50 percent by 2030, most of that in developing countries, which insisted that they did little to create the problem over the past two centuries and could not be expected to pay energy's full cost after two centuries during which we got it at a cut-rate price. They demanded that if fuel was going to get more expensive because of the environmental damage caused by our rapid industrialization, then they should be compensated for having to catch up in a world in which that damage had been done and for which they bear little responsibility. This concept, known as "differentiated responsibility," which we saw rear its head in the aviation dispute in chapter 1, runs like an unsettling current through the world's efforts to devise a price for carbon. They wanted money, not emission limits, to help make the switch to renewable energy.

The negotiations in Kyoto, as they have been in annual climate summits ever since, were like torque in a bottle. It was here that the two primary motives of the market—to raise the cost of carbon on the one hand and to keep the costs of compliance low on the other—would be set on a collision course, with a torrent of unintended consequences that we continue to live with to this day.

Patrick Birley's dominion of flickering numbers can be traced back to an afternoon in Rio de Janeiro, Brazil, in November 1997, one month before the Kyoto talks. The top three people in Brazil's climate delegation caught a morning plane for Rio out of the nation's capital, Brasilia. It was a fast, action-filled day, remembered José Domingos Miguez, the country's chief negotiator on climate change at the time. Miguez is an emphatic personality; his bushy white beard and oversized goggle glasses give him the aura of an excitable college professor. He worked as an engineer for the national oil company, Petrobras, before joining the Ministry of Science, Technology & Innovation in 1994 to coordinate Brazil's climate strategy. Brazil, with an economy bigger than Russia's, is one of the world's largest emitters of greenhouse gases, due largely to its rates of deforestation. The country is a key player in the tangle of global interests that converge around climate change. Preparing for the Kyoto negotiations to come, Brazil was also representing the G-77, the negotiating umbrella for developing countries.

The Brazilians awaited their American counterparts in the colonial-era Hotel Gloria, gleaming white with colonnades and terraces and located just blocks off Copacabana Beach. By the afternoon, the Americans had arrived—one each from the State Department, the EPA, and the office of then vice president Al Gore, who was leading the American effort. "The meeting was one day," Miguez recalled. "They flew down to Rio by lunchtime and went back to Washington in the evening."

Miguez presented the Brazilians' proposal: Industries in developed countries would pay financial penalties for their excess emissions, and a portion of those funds would be directed to help mitigate the impacts of climate change in developing countries and facilitate their switch to less fossil-fuel-intensive energy sources. Those with a "historic responsibility" for climate change, Miguez said, would pay for slowing it down. The Brazilian Proposal had the strong backing of the other major player on climate negotiations, the European Union.

To the Americans, Miguez's proposal sounded suspiciously like a tax on carbon. "Political suicide" were the words that Al Gore would later use to describe the prospect of going back to Washington with a proposal for a carbon tax. The Americans wanted to use the market to leverage business-to-business transactions through the purchase of emission allowances rather than having governments mandate a penalty for carbon emissions. This would also, the Americans insisted, take the sting out of the costs to industries of compliance with greenhouse gas restrictions—a critical factor to selling the policy back home.

In time for the return flight, as darkness fell, the Americans prevailed. "We acknowledged the political difficulties in obtaining our initial idea" is how Miguez more than a decade later would diplomatically recall the Brazilians' concession.

In Kyoto, Brazil and the United States presented their new joint plan to the world: They proposed that carbon be given a price via the supply–demand calculus of the markets. The Europeans came to the negotiations fiercely opposed to this idea; they preferred a more direct regulatory approach, a tax on carbon. "The European Commission's idea at the time was, 'If you think we're going to trade away twenty years of regulatory controls for the idea of people being able to trade away their right to pollute, its not going to happen,'" recalled Henry Derwent, who was an economist at the UK's Department for Environment, Food and Rural Affairs during the Kyoto negotiations. "They were deeply suspicious of carbon trading."

But trade it away they did, as did their Brazilian allies. After contentious negotiations, and facing the threat of American withdrawal, a resolution was reached: They would sell the air. More precisely, they would sell the right to pollute the air.

A cap on carbon emissions was established that would over time get progressively tighter. Companies would be allowed to purchase—aka trade—emission allowances if they were over the cap, and sell them if they were under it. They also agreed to the idea that companies could buy themselves out of a portion of those limits by investing in UN-certified emission reductions in a develop-ing country—giving rise to the principle of offsets (more on that in the following chapter).

Thus was born the international carbon market. "Cap and trade" became the centerpiece of the world's fight against climate change. "Amazing!" is how José Miguez, more than a decade later, would characterize what that one-day meeting in Hotel Gloria had wrought.

What it wrought was a new common wisdom, delineating the bookends of a debate that continues to this day over how to apply a price for carbon on a global scale. Everyone agreed that carbon needed a price. But the price was to be masked behind the vast amounts of capital passing between polluters and speculators. The price of carbon would gradually be nudged higher via the ebbs and flows of supply and demand, rather than being directly applied to the energy system. The world's primary climate tool, it was decided, would hinge on the short-term imperatives of companies compelled to obtain allowances to meet new emission limits.

The deal created lingering strategic divisions between large envi-ronmental groups like the Environmental Defense Fund and The Nature Conservancy—which claimed that the massive amount of money needed to dodge the worst impacts of climate change and slow it down could come only through the market-based channeling of private capital—and others like Greenpeace and Friends of the Earth, which claimed that there were more direct, transformative ways to influence companies' decisions over the use of fossil fuels.

What it also wrought was the complex Rube Goldbergian appa-ratus designed to avoid what was perceived as the politically far more difficult option of a carbon tax. And it came with another key distinction between pork bellies and carbon. While the supply of the former physically exists, the supply of the latter is determined by emission reduction targets: Carbon's value is conjured out of the gap between the emission target and the emission reality, an abstraction that the police agency INTERPOL would later label a prob-lematic "legal fiction." A new economy based on these underlying principles, representing a historic convergence of the financial and environmental communities, was poised to spread across America and around the world. Dozens of new traders set up shop. The city of New York set aside an entire floor of the New York Mercantile Exchange along the Hudson River for carbon trading.

Then in March 2001 the deal unraveled; the United States jumped ship. Shortly after taking office, the newly elected President George W. Bush withdrew the United States from the Kyoto process and yanked it from the possibility of ratification in the Senate. Bush flipped the notion of "differentiated responsibility" on its head, asserting that the United States would sign no agreement that did not also subject China to emission limits.

The floor set aside in the commodity exchange would stay vacant for the better part of a decade, and dozens of companies preparing to create and profit from the new mechanism would practically overnight go out of business. Richard Sandor, an American derivatives trader and economist, had hoped to launch a new carbon exchange in Chicago to rival the one in New York, but was forced to settle for a far smaller voluntary market, facilitating the sale of emission offsets that were subject to no independent government oversight (and underwent media scrutiny revealing numerous instances when projects produced far fewer emission reductions than promised). Others had their best-laid plans shunted aside, too: General Motors, Chevron, and American Electric Power, the three companies we encountered in chapter 4 in southeastern Brazil, created the Guaraqueçaba forest preserve because they assumed that the United States would someday establish emission limits that they could offset with their trees. When the US Congress refused to pass cap-and-trade legislation in 2009, the companies were left with offsets that had no legal standing and could only be sold in the voluntary market and/or used for public relations purposes to promote their trade-off with home-bound emissions.

After Kyoto, Henry Derwent left the UK Department for Environment, Food and Rural Affairs to become president of the lobbying organization for the world's carbon traders, the International Emissions Trading Association (IETA). Derwent recalled those tumultuous days in the wake of Bush's decision, when financial players shifted their focus across the Atlantic: "Because of the unbelievably blind policies of the US administration, we could reestablish and assert the city of London at the center of the markets . . . Wall Street, all the people who had dominated for thirty years, could not participate. And now London is the center of the fastest-growing commodity trade in the world." Europe, not the United States, became the financial pivot to the world's effort to reduce carbon emissions.

The Europeans were left to execute the approach bestowed upon them by the Americans. Richard Sandor went on in 2005 to found the European Climate Exchange, and hired Patrick Birley to run it.

The narrow, traffic-jammed streets of downtown London were soon cluttered with the offices of carbon traders. Based on the theory that liquidity in the market would help leverage the price up, the market was designed to encourage speculation. The more financial liquidity, it was presumed, the more money at play, and—assuming a growing economy and steadily lowered emission caps—the more demand for emission allowances and, at the end of that chain, a higher price for carbon. Boutique carbon project developers like EcoSecurities and Climate Change Capital were just down the block from the carbon trading desks of Goldman Sachs, Merrill Lynch, JPMorgan Chase, Barclays, Deutsche Bank, and other major European and American finance titans. By 2010, thirteen of the top twenty carbon trading firms, according to research by Friends of the Earth UK, were financial firms.[10] (Patrick Birley left the ECX after it was sold to the Intercontinental Exchange (ICE), the world's biggest commodity exchange, which sells everything from futures on oil and gold to futures on carbon.)

Many of the new players in the fields of the clean dark spread resembled the players in the fields of the old, dark spread—a sure sign that the convergence between the financial and environmental communities had taken hold. Twenty-five percent of all carbon trades, according to the World Bank, occur on speculative, secondary markets. Much like mortgage securities, these trades are not connected to the underlying item itself—carbon emission allowances—but to the price of those allowances on the primary market. Legions of new traders who'd speculated on the price of oil and gas switched to the price of the waste product from those commodities—greenhouse gases, carbon.

Sometimes it seemed as if they just bounced their conceptual carbon back and forth across the street. Marius-Christian Frunza, who traded for a French company at the height of the market and now works as head of research for the eCO2market carbon consulting firm in Paris, commented, "If I buy a BMW from you and sell you a Lincoln for the same price, that looks like an exchange of, say, one hundred thousand dollars. You have a car, I have a car. But in reality nothing economically has changed for either of us." Such transactions may make incremental profits for the financial firms that wager minute-by-minute on the price of carbon, but make little direct contribution to reducing emissions.

For the first two years, the market followed the textbook: Prices steadily rose. By the middle of 2010, the carbon price had reached thirty-two dollars a ton. Then the assumptions on which the market was based started to collapse. More precisely, the twofold goals of producing a price high enough to trigger changes in the sourcing of energy and reducing the costs of compliance collided—and the latter won. The economic recession was devastating to the hope that the familiar patterns of the market could leverage us away from fossil fuels. Production shrank across Europe, and with it the demand for energy. Companies were issued pollution goals founded on a baseline from more prosperous times in 2005, and were left with an excess of emission permits. The British NGO Sandbag— named after the barricades erected to protect against flooding and other natural disasters associated with climate change—concluded that the European Commission had issued 2.2 billion tons worth of excess emission permits. Demand for them plunged, and with it the price of carbon. In 2011, the price dropped to around twenty dollars a ton. Between 2012 and 2013, it dropped 49 percent, according to Point Carbon. For most of 2013, the cost for carbon hovered between four and eight dollars a ton.

Whichever price you choose, from Patrick Birley's low of $35 to the EPA's $38 to Nicholas Stern's $85 to Thomas Heller's hopes for $150, since peaking in 2010 the carbon price has come nowhere close to reaching any of them. And with the depressed price went any overtly economic rationale to switch away from fossil fuels. Even the Confederation of British Industries weighed in with concerns over the long-term implications of the market's plunge. The bottom-feeder prices of 2013, it stated in a submission to the government, are "not driving the low-carbon investments that are needed now to meet climate goals cost-effectively over the period to 2030 and beyond."[11] Coal use in Germany spiked, as it did in the UK and other major economies of Europe.

A report by Deutsche Bank—which provides monthly assessments of the carbon markets—shows clearly the immediate consequence of the dropping carbon price and its relationship to the quest for renewables. After an initial proposal by the European Commission to shore up the price by restricting the supply of emission allowances was blocked by the European Parliament in April 2013, the stock of RWE, the continent's major coal-fired utility, rose by 1.5 percent. In the same week, the value of stocks in Europe's biggest hydro, wind, and nuclear power generators, whose value depends partly on the price that companies like RWE have to pay for carbon, dropped by about 8 percent.[12] "Everybody knows the market is a fiction," commented Jason Anderson, a senior policy analyst at the World Wildlife Fund in Brussels. "But once it's up and running, they need to behave as if it's real. . . . The only real beneficiary of low carbon costs is the coal industry."

For polluters, there's been a serious upside: They can buy cheap allowances and bank them against the day when the price rises—thus providing little incentive for investments in energy-saving technologies. The Environmental Defense Fund estimated that the glut of emission allowances meant that airlines, still battling with the EU over their emission requirements, could purchase allowances at a rock-bottom price that would amount to less than the profit they make on selling snacks and baggage privileges.

Many companies turned the low-cost credits into a new source of profit. The huge Luxembourg steel maker ArcelorMittal, for example, when faced with declining demand for its steel in 2012, suddenly found itself with 218 million tons of excess emission allowances—which it had been allocated for free—and promptly sold them to other companies anticipating a need for more emission allowances in the future. The steel company made a hefty profit of about two hundred million dollars on the deal while demand for its steel was dropping, and did nothing to trigger a switch from the coal that the company uses to power its factories. Another example comes from Greece, where—as the economy imploded in 2012—some of the country's biggest polluters came nowhere near their allotted emission cap. The country's two largest cement companies made millions of dollars selling the emission allowances they no longer needed. "The supply of carbon allowances was intended to create scarcity against business-as-usual emissions and to encourage low-carbon investment," said Damien Morris, senior policy adviser to Sandbag, which helped out the deals. "It was not intended to reward industrial firms for lowering their [production] output." (Needless to say, the economic crisis also illustrated a fundamental fact: Lower the rates of consumption and down go the rates of production—and with them the emission of greenhouse gases.)

Nor were limitations placed on companies' ability to pass along their additional costs to consumers. In fact, European utilities were found to have been passing along added costs during the ETS's first phase of carbon trading, though they'd been issued more than 90 percent of their allowances for free during the markets' first several years.

By the end of 2013, and with the price in the doldrums, carbon traders were leaving the business in droves. Barclays, Goldman Sachs, Bank of America, and others closed or shrank their trading desks in London. Many traders switched back into traditional energy commodities, speculating once again on the elements that created the need to create a commodity out of carbon in the first place. "They're shifting chairs on the *Titanic*," said Michael Dorsey, a professor of environmental studies at Wesleyan University who has tracked the evolution of the markets since their creation. There's another reason they retreated, too: The carbon markets were turning into a high-crime zone that undermined their integrity, a phenomenon we'll explore in the following chapter.

In a frantic effort to resurrect the markets, in December 2013 the European Commission proposed once again to retain nine hundred thousand tons of allowances that were otherwise scheduled to be released onto the market in subsequent years. This time, the measure was passed by the European Parliament. Withholding one-third of the supply of new permits had the effect of slowly stanching the price collapse; by early 2014 it had crawled back to about five euros, just over eight dollars a ton. Point Carbon anticipated there would still be a significant glut of permits at least until 2020.

The roller-coaster swings in price highlight the vulnerabilities involved in subjecting the primary tool for leveraging lower emissions to the volatile mercies of the market.

Nevertheless, cap and trade is expanding in several of the world's fastest-growing economies as what is perceived to be the least disruptive and least expensive means for creating a price for greenhouse gas pollution.

In 2013, China launched carbon markets in seven of its most industrialized provinces for a select group of heavily polluting industries. Other countries, like Korea and Mexico, took their first steps toward markets, too, though they're at this stage entirely voluntary. Europe was scheduled to start the process of linking its carbon markets with those in Australia, until that country's new prime minister started backpedaling on the idea.

California's new market, fully functional as of 2014, puts the state in closer alignment on the climate with Brussels than with Washington, DC. Its system is based on principles similar to those of the European ETS—though the fact that it involves just one jurisdiction, the state, makes it simpler to oversee than the twenty-eight countries involved with the European market. It is expected to generate about ten billion dollars in transactions. (The California market is also linked to a similar market in the Canadian province of Quebec.) There are three key distinctions between the European and the California approach. First, the state set a ten-dollar-per-ton floor on the carbon price in an effort to avoid the total collapse that occurred in Europe; ironically, through much of 2014, California had the world's highest price for carbon. Second, the state's Air Resources Board closely monitored and verified emissions by the state's major industries to establish a reliable baseline; the Europeans relied on voluntary emissions reporting in its early years, which contributed to gross over-allocation of emission allowances, a discrepancy that continues to haunt the market. Third, California permits those avoided deforestation offsets we visited in chapter 4, while Europe excludes them—presenting a potentially serious obstacle to the state's hope of linking its trading system to that of Europe by 2020, which will require establishing compatible mechanisms governing offsets and emission reduction targets.

Another market experiment, called the Regional Greenhouse Gas Initiative, or RGGI, has been unfolding in nine New England and mid-Atlantic states, and targets only utilities (while California's measure targets greenhouse-gas-intensive industries as well as electricity generators). Since 2008, RGGI has generated more than a billion dollars from the sale of emission allowances, funds that for the most part were channeled into renewable energy and energy-efficiency programs and in rebates to consumers.

In the middle of 2014, Obama began to try to recalibrate the American role in the global effort to reduce emissions. His EPA issued new guidelines governing the greenhouse gas emissions from coal-fired power plants, aimed at reducing emissions 30 percent from 2005 levels by 2030. To accomplish that, the EPA issued emission-reduction guidelines to every state—which are expected to give rise to a further boom in statewide or regional carbon markets over the coming decade—for which California and RGGI are presumed to be models.

The ultimate aim is to give that one-ton balloon of CO_2 a uniform price through a globally synchronized market. In the dream of the market's architects, emission allowances and offsets for aviation, manufacturing, oil refining, and food transport could be purchased in markets across the globe.

Watching the numbers flicker on Birley's screen, I breathed the air of England. According to the principles of the carbon markets, that air has the same value as the air anywhere else in the world. By the same token, the carbon markets have the unanticipated side effect of illustrating just how shared this atmosphere actually is by us all. Those flickering numbers seemed to run like a tab of the world's emissions, each new digit a signal of what we're willing to pay to preserve the atmospheric balance and pursue less-destructive sources of energy.

But the functioning of the carbon markets thus far raises the question: what is it, exactly, that's being paid for? The markets were the preferred alternative of business, so it's worth assessing their performance according to traditional business criteria: are they the most cost-efficient way to propel us toward a future less tied to fossil fuels and the destructive impacts that come along with them?

A commodity market has one fundamental requirement: reliable measuring of the commodity itself (the way we would weigh, say, the belly of a pig, or, for that matter, an ounce of silver). Turning the world's foremost pollutant into a commodity based on an abstraction, subject to the same vagaries of the market as any other commodity, comes with a unique set of enforcement challenges to ensure that one abstraction is not simply exchanged for another. Carbon the gas is a highly unstable member of the periodic table; its molecules cling to one another in the atmosphere, which is why they take so long to break down. Carbon the commodity, however, is elusive, given form depending on the needs of those who create it, malleable and difficult to measure and track precisely because of the process by which a very real family of gases is conjured into an abstraction with value. This creates a set of vulnerabilities that, in addition to the external volatility we've seen, also come from within the market itself, in which its unique characteristics have made it vulnerable to manipulation—of both the illegal and the perfectly legal variety.

CHAPTER 8

Trading in Hot Air

Stealing Carbon

O n the morning of January 18, 2011, Czech police received an alarming telephone call. A bomb had been placed in a building on Sokolovskaya Street, in a commercial district just blocks off the Charles River in downtown Prague. Most of those inside worked at the offices of the OTE, the government-owned Czech Carbon Registry. The bomb squad arrived and evacuated the building. Barricades were hastily assembled, and trading was shut down during the prime morning hours. A hazmat team combed the corridors with explosives-sniffing dogs.

The team found no bomb. A day later the police identified what had actually happened. While the carbon exchange was being evacuated, a gang of cyber-thieves had invaded computers on the trading floor and looted the commodity nested inside them. Whoosh, gone: With the flick of a finger on a keyboard, more than two million tons of greenhouse gas emission allowances were stolen and sent into the criminal underground. The Czech version of the ECX had been hacked of two million balloons' worth of CO_2.

Police speculated that the telephoned bomb scare was a feint to divert traders from noticing exotic cursors moving across their screens. It was as if thieves had robbed a bank—except the bank held the permits that Czech companies are supposed to purchase to emit greenhouse gases; and the thieves used other computers, not guns, to steal the serialized digits—the "currency"—on traders' screens. Each permit to pollute one million tons of greenhouse gases is the equivalent of money. At the time, a permit was selling for about twenty dollars a ton. Within a week, the missing pollution permits had been traced to exchanges in Estonia, Poland, and, finally, the BlueNext carbon exchange in Paris. Europol estimated that the thieves made off with about thirty-eight million dollars from black-market sales to companies that purchased the pilfered credits.

This was not the first or the last example of how the particular characteristics of carbon make it uniquely vulnerable to criminal infiltration. As a financial vehicle, carbon has an elusive personality. It's precisely because you can't hold carbon in your hand—as you might a pork belly or for that matter an ounce of gold—that the international police agency INTERPOL subsequently termed the market in carbon a "legal fiction," because of the counterintuitive nature of turning a gas created on every corner of the planet into a commodity in order to make it disappear. Criminals introduced a new chapter into the annals of white-collar crime, taking advantage of the difficulty in regulating the abstraction that's at the heart of the carbon market.[1]

Six months before the Czech theft, an enterprising Danish journalist, Bo Elkjaer at the newspaper *Ekstra Bladet*, revealed that 80 percent of the carbon trading firms registered in Denmark were in fact criminal fronts engaged in an elaborate tax scam. They charged unsuspecting buyers of emission credits the customary 25 percent value-added tax on the transactions—then pocketed the extra funds and skipped town. The scam was apparently under way during the Danish-hosted climate negotiations in Copenhagen in December 2009. In a rush to kick-start the market, the Danish authorities had neglected to ask for verifiable IDs or other proof of traders' integrity, as they do on other financial markets. Similar schemes were unearthed in the Netherlands, Germany, the UK, and Norway. When the authorities finally closed the loophole that facilitated the scheme, the volume of carbon trading in those countries plunged by 90 percent. The network of illicit traders deprived European authorities of at least five billion euros, around seven billion dollars, at least some of which was supposed to be used to support climate mitigation and renewable technologies. The international police agency INTERPOL warned European governments of the vulnerability of the carbon markets to infiltration by criminals.

Inside the cavernous atrium of marble and stone at INTERPOL's headquarters, a heap of discarded computer parts lays sprawled across the floor. The pile of cracked motherboards, fractured screens, and rotted wires looks like an art installation—a strange site inside the heavily fortified building of glass, granite and concrete—but it's actually a pile of inspiration. Emile Lindemulder, an agent with the pollution crime unit of INTERPOL's Environmental Crime Programme, explained to me that the e-waste serves as a reminder of one of the agency's chief responsibilities, enforcing the Basel Convention that restricts the international trade in hazardous wastes.

If international e-waste dumping was the new crime on the block in decades past, carbon crime is the latest addition. In the fall of 2010, I attended a historic gathering at INTERPOL's compound along the Rhone River in Lyon, France.[2] For the first time, law enforcement officers primarily responsible for environmental violations in Africa, Asia, and Latin America—the illegal trade in wildlife, logging, and hazardous waste—were brought together with police specialists in fraud and financial crimes. Some two hundred law-enforcement participants from thirty countries included agents from the EPA and the Commodity Futures Trading Commission (which would regulate a US national carbon market if there ever is one), and their counterparts from Britain, Germany, the Netherlands, and elsewhere.

There, regulators and law enforcement were presented with a commodity they barely understood. Its definition was an entirely new creature in the commodity business, and the authorities had no way to protect against fraud they could not imagine. "When there's this amount of money involved, criminals get interested," said Lindemulder. "The carbon markets involve so many parties, so many new instruments and forms of vulnerability that we haven't been aware of before." That was still in the early stages of the evolving market; members of law enforcement, like the rest of us, were still learning.

As if to illustrate Lindemulder's point, the Czech scheme unfolded three months after the INTERPOL conference. Nine months after that, the Spanish carbon trading exchange was forced to close for several days after Spain discovered it, too, had been looted of several million tons' worth of emission allowances.

In the years following, British investigators would shut down several carbon boiler room operations in London, charging more than two dozen traders with selling millions of dollars' worth of carbon credits that didn't exist. The Paris exchange, BlueNext, was accused of handling millions of tons' worth of double trades—selling the same emission allowances multiple times. The Hungarian government was discovered to have done some sleight of hand on its own emission registry, exchanging credits it had been granted for free—due its status as a formerly socialist country that had joined the European Union—with those its industries were supposed to purchase. Carbon prices plunged.

And then there were some truly audacious schemes in the practically unregulated voluntary market, in which millions of tons of ostensible credits were put up for sale based on preservation of forests in Liberia, Tanzania, Brazil, and other countries—lands to which, it was discovered, carbon project developers did not even have legal access. Such schemes began as early as 2008 and continued into 2012, when another was unearthed in Brazil, this one involving two million acres of forest in the Amazon, and another in 2013 in the Peruvian Amazon.

Peter Younger, a senior agent with INTERPOL who helped investigate some of those cases as the agent responsible for enforcing laws protecting forests and endangered species in Africa, explained that in countries where land ownership is often disputed, the possibility for fraud is considerable. "You're obtaining not a physical entity or asset but a piece of paper. . . . In effect, you could be falsifying ownership in something you can see in order to sell something that you can't. And then inserting that into the carbon markets and selling it to people."

By August 2013, INTERPOL's Environmental Crime Programme issued a report outlining the agency's continuing concerns about the integrity of the market and its vulnerabilities to criminal manipulation.[3] "Unlike traditional commodities," INTERPOL reported, "which at some time during the course of their market exchange must be physically delivered to someone, carbon credits do not represent a physical commodity but instead have been described as a *legal fiction* that is poorly understood by many sellers, buyers and traders. This lack of understanding makes carbon trading particularly vulnerable to fraud and other illegal activity. Carbon markets, like other financial markets, are also at risk of exploitation by criminals due to the large amount of money invested, the immaturity of the regulations and lack of oversight and transparency."

The crimes set loose a domino effect, creating unease among traders and speculators over the integrity of the markets. Criminal activity depressed a price that was already on its way down due to the surplus of emission allowances caused by the global recession and other factors. Investors retreated, the price declined further, and as the price declined, more investors retreated, a deadly cycle. Marius-Christian Frunza, the former carbon trader now working with the French carbon consulting firm eCO2market, explained that the legal violations of the market presented an existential threat to the market's rationale. "Criminal activity," he said, "drove away investors, leading to a decline in price. And the money they took was from our governments, from our pockets." The money they took was also from our pool of collective hope that subjecting carbon to the vagaries of the market would ultimately provoke a significant shift away from fossil fuels.

What has happened in the carbon markets is at one level strangely routine; they're like a microcosm of the broader financial markets. Various versions of these schemes have been used against almost all the various commodity markets, from silver to pork bellies. As Frunza put it: "In the equity markets, you had fake IPOs, fake pump-and-dump schemes, you had bankers that tried to manipulate the market with mortgage derivatives and other devices. . . . In carbon [trading] you had all those kinds of schemes, characters who act like street crooks and wise guys."

On a far more profound level, however, these are not your typical white-collar crimes, for the carbon markets were created not just to sell the carbon equivalent of pork bellies; they were created with a higher purpose in mind. It is this crosscurrent of motivations, to find a way to both generate profits in the traditional way and reduce fossil fuel energy consumption in a new way, that has been the markets' most dangerous flaw.

———

Criminals are not the only threat to the integrity of the carbon markets. Another threat—perfectly legal but involving a magnitude of difference in its implications—comes embedded in the system itself. That threat can be found in the quite elegant equation of trading emissions in one place for emission reductions somewhere else—in other words, offsets.

Negotiators in Kyoto gave the United Nations the unprecedented job of inventing, facilitating, and overseeing the multibillion-dollar market in offset transactions between rich countries and poor ones. Never before has the UN been asked to be a financial regulator.

At the core of the offset transaction is a novel principle known in UN lingo as "additionality." The idea is worthy of conceptual art. Carbon project developers must demonstrate to the executive board of the UN's Framework Convention on Climate Change (UNFCCC) a counterfactual: Funds from polluters will trigger additional emission reduction initiatives that would not have happened without their financial support. That commitment, once approved by the UN, is then is turned into a commodity for sale on the international carbon markets. Allowances, like those stolen from the Czech registry, are permissions to continue polluting in one-million-ton increments. Offsets, which cost generally 50 to 90 percent less than allowances, are sold in one-million-ton bundles, also on the European Climate Exchange. They either are a UN-approved commitment not to pollute in one arena, in order to continue polluting in another, or represent participation in a project to soak up existing levels of CO_2 (in the case of planting or conserving trees) over a time period often measured in decades. Companies can use offsets to meet as much as a quarter of their total emission obligations.

One of the first companies in the world to give offsets legal definition is a Salt Lake City–based firm called Blue Source. In the late 1990s, the company, like many in the United States, anticipated a boom in environmental markets in the wake of the Kyoto accord. Blue Source at the time specialized in building pipes for oil companies to store and transport excess CO_2. Bill Townsend, the company's CEO, told me that after Kyoto, he and his business partner realized that rather than providing this service to oil companies, they could switch gears and develop new ways to sequester CO_2 rather than transport it—and sell that service as an offset in accord with the newly affirmed environmental goals. Then they faced a significant question: What was it, exactly, that they would supply?

"There were two of us with a couple of blank sheets of paper," Townsend recalled. "We're thinking: *What are we selling? We're not selling the commodity, carbon* . . . We had to create the word in order to have a transaction." Eventually the two pinpointed their product: "It was not the carbon but its elimination," Townsend explained. "We're selling the right to market the elimination of carbon." They hired a lawyer and drew up a legal definition. The United Nations would make a similar conceptual leap as it set up an entire office at the UNFCCC to administer what would become a multibillion-dollar market for the buying of offsets in developing countries.

The program, an outgrowth of the original Brazilian proposal for a fund to help developing countries adapt to climate change, was given a very un-market-like name, the Clean Development Mechanism, or CDM. Entire new professions were invented—carbon accountants, carbon auditors, carbon counters—to shuttle to the CDM what has amounted to about one offset proposal a day from 2008 to 2014.

Fifty-two offset protocols—standards for offset projects—have been approved by the UN, and approximately four thousand projects were certified using those protocols over that period. A greenhouse gas emitter can get credit for helping a Mexican pig farmer install a device for capturing methane from livestock waste; help an Indonesian factory relying on coal transition to a cleaner fuel; build a wind farm in the open plains of China; plant new trees in Brazil or Papua New Guinea on land degraded by deforestation; or build a dam in India to channel water into a turbine to generate energy, on the presumption that otherwise that energy would come from coal.

And then there's coal, which is responsible for a quarter of all the world's greenhouse gases. Six coal refineries in China and India were certified in 2013 to kick off seven million tons of offset credits based on claims that they would install new super-critical combustion processes that emit less CO_2. The irony of giving financial support to the world's single most significant contributor to climate change, an initiative that has been promoted aggressively by the global coal industry and was finally successful in 2013, highlights one of the central challenges to the offset system. Do we, as the Center for International Environmental Law put it during a furious protest campaign against the coal offsets during the climate negotiations in Warsaw, use funds that are supposed to shift us away from fossil fuels to spur incrementally cleaner coal refining and thereby "lock in millions of tons of CO_2 emissions over decades to come"?[4]

International offsets have been a backhanded way of addressing the large gap in responsibility between developed and developing countries for the greenhouse gases already in the atmosphere. In total, more than twenty billion dollars have been channeled into corners of the developing world that would have been far off the beaten path of investment capital were it not for companies investing in offsets to maximize their allowances. "In economic terms, the CDM has been a transfer of wealth from the north to the south," commented Blas Pérez Henríquez, an economist and director of the Center for Environmental Public Policy at the University of California, Berkeley.

But as long overdue as such transfers of wealth may be, there are major questions as to whether this unprecedented north–south flow of cash is actually resulting in a one-to-one relationship between emissions in one place and emission reductions somewhere else. The fundamental challenge revolves around measuring additionality, the counterfactual claim to not emit in the future.

To make that measurement, the UN certified some two dozen companies to validate emission reduction claims, and then verify, often years later, that those reductions have occurred. These companies are known as "validators"—the carbon equivalent of, say, Moody's, which rates the strength of bond offerings. Most have roots in the maritime, insurance, and accounting businesses, and among them are the accounting firm Deloitte Touche Tohmatsu, the British transportation safety firm Lloyd's Register, the Norwegian maritime inspection service Det Norske Veritas, DNV, and the SGS Group, founded in France more than a century ago to verify the weight of grains traded across Europe. Now all of these companies are supposed to verify the weight of CO_2 not emitted by offset projects as emerging economies race forward with fossil-fuel-heavy development strategies.[5]

Questions, however, about the measurements abound. The essence of the problem: The validators are chosen and paid for by the carbon project developers themselves, who sometimes pit them into bidding wars against one another. On top of that, carbon markets have not been subject to the rules devised in response to the Enron scandal, in which accounting firms were compelled to distance themselves from consulting on financial strategies. So the same institutions that audit claims by emission reduction project developers can also help strategize new projects.

As a result, some of the biggest carbon financiers have been permitted to develop and sell their own offset projects, providing a vested interest in the results of the additionality tests. JPMorgan Chase, in the most striking example, paid two hundred million dollars to purchase the world's biggest carbon project developer, EcoSecurities. The deal offered synergy, with a climate twist: The finance giant vertically integrates the offset business, offering the opportunity to provide seed money to develop carbon offset projects, pay for the validations of their potential for emission reductions, and then profit from the offset funds generated in the carbon markets. In the United States, Goldman Sachs purchased a 50 percent interest in Blue Source, the Salt Lake City company that was in on the ground floor of the offset business; without a federal-level cap-and-trade system, though, the company has limited itself to the far less lucrative voluntary carbon market, where it's now the biggest developer of forest offsets in North America.

Against this backdrop, flawed accounting abounds. Just as the housing bubble was created by pooling mortgages into derivatives, a carbon-value bubble has been created by bundling emission reductions of varying quality. And just as Arthur Andersen inflated the value of Enron before its collapse, the carbon market's accountants operate within a self-contained circle. Efforts within the UN to establish a more independent auditing system—implementing fixed fees or direct payments to the UN, for instance—have been repeatedly beaten back by the validating companies allied with many of the countries that have been the biggest beneficiaries of offset funds.

Not surprisingly, the nexus between offset bundling, auditing, and selling attracted the attention of INTERPOL, which described, in its 2013 report on the criminal vulnerabilities in carbon trade, the system's "inherent conflict of interest in which the [auditors] are incentivized to facilitate the project's approval rather than to assure accuracy of the validation process."

In 2008 and 2009, the two biggest auditors, DNV and SGS, were temporarily suspended by the UN when their calculations of additionality on numerous projects were found to be flawed. Together the two companies were responsible for affirming the emission reduction claims of more than two-thirds of the offset projects approved in the first two years of the program. But the UN still does not have the power to withdraw already approved offsets, even those of dubious origin. Forcing companies to obtain new offsets to replace questionable ones was found to be too risky, according to Clare Breidenich, a former climate policy analyst at the US State Department who later ran the UN department monitoring developed countries' greenhouse gas emissions. "They were afraid," recalled Breidenich, "that if that were the case, there would be no market." The interest in generating market liquidity trumped concerns over offset quality.

Rising concerns about the integrity of offsets prompted the CDM Policy Review Panel, a group of experts convened by the UN, to review more than a thousand offsets approved by SGS, DNV, and hundreds of others.[6] The panel parsed through thousands of pages of offset proposals and validation documents submitted through 2012. They came to an astonishing, though little-noticed, conclusion that rocks our understanding of the attempt to create a global balance of emission trade-offs. "[A] substantial portion of these projects," they said, "should be considered non-additional, leading to a significant net increase in global GHG emissions."

Translation: The offset system is likely leading to a net *increase* in emissions, according to a body of experts assembled by the very same body that facilitates those transactions. The only reason for a company, say in Germany, to purchase an offset is that it wants to keep emitting and finds it cheaper to buy an offset than, say, to buy an allowance or invest in new technology that might reduce emissions. But if offsets are not additional, there's no real trade-off; the offset provides no real balance to the very real emissions back in Germany. Industries keep emitting without any distant offset to balance them out.

The panel identified most of the non-additional projects as large-scale dams, wind farms, and those Chinese and Indian coal refinery offsets—which together, if current policies continue, are destined to constitute at least half of all certified emission reductions by 2020. Claims of additionality on such projects, panel members said, are "not fully credible" due to the fact that such initiatives provide long-run cost savings to their owners, and thus have no need of CDM funds. "The chance they wouldn't happen without the carbon market is very unlikely," commented Lambert Schneider, a German environmental engineer who contributed to the report and serves on a UN advisory board on offset methodologies.

Schneider asserted that the financial risks involved with pursuing a project without CDM funds were often exaggerated, and that claims of additionality in the proposed coal projects were "not credible." Such projects, he said, have their own underlying financial logic independent of the CDM; saving energy is good for business, and generating energy with hydropower is a long-term and worthwhile business investment with the potential to generate profits for decades. Most of the CDM wind projects are in China, which already has a web of subsidies, tariffs, and tax breaks to support further development of the technology; the country is now the world's leading producer of wind turbines.

At least 40 percent of the offset projects reviewed by the panel, Schneider told me, would have happened anyway. This means that a minimum of four out of every ten dollars spent on international offsets are likely contributing little, if anything, to slowing down the onrush of greenhouse gases into the atmosphere. Offset funds just offer the crème on the top of existing profit margins. Schneider proposes eliminating half of the existing cache of offsets to encourage only offsets that would not have happened anyway and ensure that emission reductions are actually delivered.

The most acute and costly example of lax controls in the CDM, among many, comes from China, recipient of more than two-thirds of all offset funds. It took an investigation by two European NGOs, the Environmental Investigation Agency and Carbon Market Watch, to reveal how those funds had actually been used. More than half were channeled at drastically excessive prices for the destruction of a single industrial gas, hfc23, a waste product from factories producing refrigeration equipment that is ten thousand times more potent a greenhouse gas than CO_2. The practice triggered outrage in the European Parliament. "European companies are spending a billion euros to buy something that costs less than one hundred million euros," said Theodoros Skylakakis, a Greek member of the European Parliamentary committee that investigated the hfc23 offsets. He claimed that it was little more than a costly effort to sustain Chinese support for the UN program, a gift of billions of euros wrapped in concerns over climate change.[7]

The European Commission agreed to withdraw hfc23 offsets in the middle of 2013, but not before we obtained a glimpse into the ironic crosscurrents of global climate policy. The Chinese government recognized the windfall at work, and slapped a 65 percent tax on profits from the hfc23 emission reduction schemes. Many of those hundreds of millions of dollars went to support the development of China's wind and solar energy industries—which enabled them to compete aggressively with European and American manufacturers. From a climate perspective, the dramatic expansion of China's renewable energy capacity should be celebrated. Instead it triggered a series of trade disputes, in which the Chinese were accused of using those and other subsidies to undercut European and American wind and solar producers in the global marketplace.

The afflictions that have weakened the offset market are eerily similar to those that have rampaged through the traditional markets. The difference, however, is that there is no threat of legal action, other than for clearly illegal acts of fraud or theft. There is no Commodity Futures Trading Commission (CFTC) to govern the integrity of the carbon markets. Instead, with carbon the UN serves as both the equivalent of the CFTC and commodity exchange, and is reliant on funds from participating governments, some of which have little interest in a tighter system that would deliver more reliable offsets on the other end of the great emissions balancing act.

———

I asked someone who was deeply involved in the application of the rules governing offsets to explain how such egregious shortfalls could emerge from an institution, the UNFCCC, which has been the leader of the world in sounding the alarm on climate change. Mark Trexler has been involved in the evolution of environmental markets since the early 1990s. He's worked as a policy adviser to politicians, as a carbon project developer, and as an advocate of the carbon market approach during the Kyoto negotiations; currently, as founding director of the Portland-based consultancy, The Climatographers, he offers his services to businesses trying to assess their environmental risks. Between 2008 and 2012, he worked as an adviser to DNV, one of the biggest validation firms that was suspended for inadequate audits back in 2009. Trexler used an interesting analogy to explain the flaws in the offset rules: He compared the additionality tests used by auditing firms to home pregnancy tests. Those pregnancy tests, the results of which are of great significance for every individual who uses them, are known for skewing toward false positives, of far less portentous impact than the reverse. Additionality tests are known for skewing toward false positives, too.

Tightening the offset standards to reduce the number of falsely additional projects, he said, "would result in some legitimate projects being kept out. . . . Striking the balance between the number of false negatives and false positives is a political decision, not a technical decision." Like those pregnancy tests, Trexler added, the CDM is still kicking out an abundance of false positives. While the former are, of course, of immense individual significance, the latter are of immense significance to the planet.

In other words, error is built into the system. The political decision was made—primarily by the US and European governments, under pressure from industry—to keep the supply of offsets abundant and the price low, and to live with questions concerning the veracity of claimed emission reductions. Once the money started flowing, recipients of the funds had an interest in maintaining the flow. In many ways, the principles central to the CDM were deemed too important to climate negotiators, and the goal of luring developing countries into a global deal too significant, to permit a seemingly technical matter of providing reliable controls over the offsets to derail the deal. Therein lies the perfectly legal crime that carries far greater implications than the illegal schemes that have undermined the carbon markets.

In any other market, such findings would be a scandal. With no other commodity would a transaction between something real on one end—emissions—and something significantly imaginary on the other end—emission reductions—be permissible. It's as if the Securities and Exchange Commission suddenly decided that artificially inflating the value of stocks was perfectly legal. The price paid for offsets bears little relation to what's delivered on the other end. They've been a low-cost diversion from making the more fundamental step of relying on less destructive forms of energy.

In the end, the money we're talking about is not your regular run-of-the-mill cash. Every euro (trades are conducted for the most part in euros) plowed into the carbon markets that is not resulting in emission reductions due to inefficiency of the system, conflicts of interest, or criminal activity is drawn from a limited pool of capital that is in many ways all of our precious capital, intended to help us navigate away from a potentially cataclysmic 3.5-degree Celsius rise in the atmospheric temperature. It's money diverted from the central economic challenge of our time.

The hundreds of billions of dollars necessary for mitigating greenhouse gases and adapting to and recovering from their already destructive consequences have to come from somewhere. Despite the abuses, and the loopholes you could drive a coal train through, most of those who cooperated to set up the cap-and-trade system, and the markets it spawned, believed they were creating something historic—that it could be a relatively painless way to leverage a price for carbon and thereby spur forward the transformation away from fossil fuels. Now is the time to assess the success of the markets on their own terms—as markets.

The mantra has been repeated many times over: carbon markets are a "win-win." You could *"Make money and save the planet."* It may indeed be possible to do both, but there are serious questions about whether the carbon markets as they're currently designed have the capacity to sufficiently penalize greenhouse gases enough to trigger large-scale investment shifts away from fossil fuels.

Market gospel held that the call and response of supply and demand would bring us a progressively higher price for carbon, and leverage developing countries into a low-carbon future. Much like the banks that at the height of the financial crisis were deemed "too big to fail," it was hoped that the carbon market would become so big, the amounts of money so significant, and the insertion of a carbon price into the global order of such transformational impact that the kinks in its founding assumptions could be worked out. Now we've had seven years of experience with that experiment.

The results are coming into focus. Europe appears to be on target to meet its targets of reducing emissions twenty percent from 1990 levels by 2050—by roughly twenty percent since 2005—but most of that, say analysts, are due to the shrunken economy, not to pressure emanating from the carbon market. Companies are not getting the emission reductions they're paying for; governments' claims of meeting emission targets are called into question when placed against the grid of faulty offsets. Erratic prices means it's difficult for businesses to plan for a low-carbon future. And the market has not yet come close to generating any price that would leverage a significant shift away from fossil fuels.

The question remains: How do we assign a price for carbon that reflects its actual costs? And, critically, if energy is going to get more expensive, who is going to pay to cover the increase?

The question at stake is an uncomfortable one: how do we make energy more expensive? The carbon markets were created by politicians trying to take a radical step—pricing the externalities of energy—in a very constrained political environment, defined largely by relentless pressure from the most impacted fossil-fuel industries. The challenge they were created to address remains—and responses to it are already redrawing the maps of geopolitical power and influence.

CHAPTER 9

The Coffee and the Cup

Shuffling the Decks in the New Carbon Economy

*I*t was in the city of Brasilia, the capital of Brazil, where I discovered the real price of a cup of coffee—the cup, not the coffee—as well as the meaning of a price for carbon.

Brasilia is a city of two million people dropped in utopian style into the vast flatlands of Brazil's central savanna. One of the world's most noted architects, Oscar Niemeier, carved a grid out of the plains following the basic design of an airplane—a twentieth-century symbol of modernism and the triumph of technology.

When I caught a cab from the hotel district, I found myself heading north on the "fuselage," a strip of landscaped savanna that runs along a six-lane thoroughfare bisecting the city, leaving behind my hotel in the tourist district to the south. To the west: the entertainment district, a hive of clubs, cafés, and restaurants arrayed along perfect rectangular lines. To the east: the residential zone, rapidly losing its rectangular order as the city's population soars. Finally, I landed in the "cockpit," several square blocks of government offices and the two houses of the Brazilian legislature, perched like inverted flying saucers in Niemeier's distinctive futuristic style.

Across the street from a small park with trees that showered yellow blossoms onto the ground was the national headquarters of IBAMA, Brazil's environmental agency—where in another wing I'd

met those twenty-somethings transmitting satellite images of deforestation to the police in Paraná described in chapter 4. I sat down
for an appointment with Branca Bastos Americano, the Secretary
for Climate Change and Environmental Quality in the Brazilian
Ministry of Environment under President Luiz Inácio Lula da Silva,
and one of the government's lead climate negotiators. Her assistant
brought us cups of coffee.

Like other professionals in the city—Brasilia is a company town,
like Washington, DC, filled with people either in the government,
in the opposition to the government, or in NGOs or lobby groups
pressuring the government—Americano is well aware she is living
in one of the fastest-growing economies in the world. "This is
not your old Brazil," she said. "We're serious about enforcing
environmental laws, laying out strategies to balance enforcement
with development." Since Lula's election in 2003 (he served until
2011), social welfare policies had lifted some ten million people
out of poverty and the nation had developed one of the greenest
energy grids on the planet. Today Brazil generates more than 80
percent of its energy from renewable sources—hydro, thermal,
and wind power. Only about 15 percent comes from coal and
other fossil fuels.

I asked Americano how Brazil would fare in a world in which
carbon has a price. She pointed to the ceramic coffee cup I was
holding: "See that cup. Brazil beats any country in a world in which
carbon has a price." She smiled, briefly: "Including China!" I was
startled by her quick response. "The energy used to process the
clay for that cup was obtained from water-powered hydro dams in
the Amazon. The cup was manufactured in a ceramic factory that
is fueled with biomass. It was transported here by a truck using
biodiesel. We will beat China every time in a world in which carbon
has a price."

The cup she referred to was nothing special—white, no special
designs or logo, just big enough to hold a demitasse of espresso. Very
much, in fact, like the cups I use to drink my own coffee—except

most of mine are actually made in China, and this one was made in Brazil, in the industrial zone outside São Paulo. It is, indeed, a tiny bit more expensive than the cheap coffee cups imported from China, from factories most likely powered by coal, that can be purchased at shops across the country. But the idea that this plain white Brazilian cup would become comparatively less expensive than its imported Chinese counterpart if the latter had to include the price of the energy used to make it seemed to encapsulate the central financial question posed by the climate conundrum: How does the coffee cup made with more renewable energy become at least equally competitive with cups made from more destructive sources of energy? This quandary of the cup is a tiny microcosm of the central challenge that has bedeviled the world for two decades. Responses to this challenge are breaking out all over—and sometimes in the most unexpected of places.

Three years later, on an afternoon in late August 2013 in Rio de Janeiro, a group of Brazilian, Latin American, and other developing country officials gathered for a climate conference co-sponsored by the World Bank and the state of Rio de Janeiro.

Wu Delin, the vice deputy mayor of the Chinese city of Shenzhen, rose to give a talk that rocked the proceedings. Shenzhen is a modern industrial city of eight million people in Guangdong province, the same province with which Pittsburgh had traded its burden of emissions. From the dais, Wu announced that the government in Beijing had decided it was time to start charging for the use of fossil fuels, and had asked Shenzhen and other of the country's most industrialized provinces to begin the process of creating their own cap-and-trade system. In a PowerPoint, Wu went step by step through Shenzhen's plan to be the first to require the most fossil-fuel-intensive industries to purchase allowances for their greenhouse gas emissions. His city was positioning itself to be the model, Wu said, a test run for a national program. Two hundred of the province's largest emitters—utilities, steel, iron, and cement manufacturers, all heavy users of coal—would in just four months from his presentation be subject to emission caps, and they would be expected to buy allowances on the new carbon market being created in Shenzhen. Ninety-five percent of the allowances would be issued for free in the first year, a number that would over time decline. The aim was to reduce the carbon intensity of Guangdong industry by 25 percent by 2015.

Response among the Brazilians was astonishment. "There was silence," recalled Walter Figueiredo De Simoni, the secretary of environment for the state of Rio de Janeiro, who helped organize the gathering. "We were stunned. Our response was, 'Wow! Just like that they're going to have a carbon price.'"

De Simoni, an economist by training, had spent the previous year negotiating with businesses in the state of Rio de Janeiro to kick-start a market or implement a minimal carbon tax. But he'd been foiled by industry opposition. Businesses claimed that such a move would put them at a competitive disadvantage with their global competitors, namely China. Now China was announcing it would unilaterally accomplish what De Simoni had been trying to do unsuccessfully for more than a year. "You look at the two countries," he said, still bristling at the irony several months later. "Brazil is seen as the greener one, but we're not as prepared to act. China is seen as the dirtier one, yet they are preparing much more aggressively for this greener economy." Brazil, the "environmental powerhouse," blessed with an abundance of green resources—water, biofuels, sun—had been upstaged by China, long seen as the global villain of climate change.

It was a remarkable moment in the parallel world evolving between two of the most important countries in the climate dynamic. Other markets were launched in 2014, in Shanghai, Beijing, Chongqing, and Tianjin provinces. The world's biggest manufacturer, exporter, and user of fossil fuels was beginning, at least, to give a price to carbon.

Of equal significance, for the first time industries in these provinces will have to keep a running inventory of their greenhouse gas emissions—though the data is not available to the public. Practically overnight, the Chinese carbon markets became the second largest in the world after the European Trading System.

That Chinese cup for sale in Brazil would indeed, over the years ahead, come with a small (very small) incremental price for carbon built into its cheap porcelain. Soon enough, of course, Americans would also be paying that small carbon price, as would every other country importing goods from those export centers in China. With one move, China had jumped the entrenched distinction between the developed and developing countries, creating a price that came not from a global accord or trade deal, but direct from China itself. And it was delivered to an audience in a country that is often an ally in climate negotiations, and always a major market and sometime global competitor for China.

The Brazilian–Chinese axis is one rich with kinetic power. Two of the most dynamic and fastest-growing economies in the world are navigating the wobbly new frontier wrought by climate change, on the tightrope between rapid economic growth and global pressure to restrain their reliance on fossil fuels. China, which relies on fossil fuels (mostly coal) for 90 percent of its energy grid and generates most of its emissions through industrial production, is like a dystopic inversion of Brazil, which relies on fossil fuels for just 15 percent of its energy grid and kicks off most of its emissions in dead trees. Brazil has been an environmental leader since the first climate change conference was held in Rio in 1992. That conference laid the groundwork for the Kyoto Protocol five years later, though Brazil has most recently been backpedaling, tilting toward pursuing a massive network of dams in the Amazon that require felling huge swaths of the forest in the quest for more water power and rapid economic development of the country's poorest states. China, the world's leading emitter and biggest user of coal, is a newcomer to the green stage, but has already become an environmental powerhouse in its own right. It is the leading producer of solar panels and a major producer of wind turbines and technology. Despite their different approaches, both China and Brazil have high stakes in the rising acknowledgment of the costs of carbon—a wild card that could speed or slow their steady ascents on the world economic ladders.

Brazil commissioned its own "mini Stern Report" from a team of economists at the University of São Paulo, who concluded that by 2050 current climate trends could cost the Brazilian economy between five hundred billion and two trillion dollars, the latter number of which is roughly equivalent to "wasting at least one whole year of growth over the coming 40 years."[1] Recent conditions in Brazil have unfolded like a sequence of costs foretold. Droughts in 2012 and 2013 left parts of the northern Amazon unusually parched; at the same time, extreme storms farther south and east, a zone that had previously been considered hurricane-free, were hitting coastal areas with a violence like never before. Not long after I visited the town of Antonina, in the state of Paraná just outside the Guaraqueçaba carbon reserve, an extraordinary downpour of rain and intense winds led to avalanches from the surrounding hills, which inundated much of the lovely town with mud, and crushed a local highway. A similar disaster occurred in the following year in the municipality of Santa Catarina, built along the hills that ring Rio de Janeiro. These extreme events (there have been others) were attributed widely to the disruptions of climate change. The tens of millions of dollars in rebuilding costs from those two devastating events alone, commented Manyu Chang, who worked in the Forestry Department of Parana state during the landslide and is now researching Brazilian vulner-abilities to climate change at the Federal University of Rio de Janeiro, "were twenty times more than the cost of adapting now and start-ing to reduce greenhouse gases." Researchers at Oxford University predicted that the area for cultivation of Brazil's major crops could also shrink significantly by 2050 under current trajectories—for soybeans, the country's dominant crop, by as much as 24 percent; for rice by 4 percent; and for corn by 12 percent.[2]

In China, the announcement of the country's new carbon markets came three months after the Chinese Academy for Environmental Planning, a scientific arm of the government environment ministry, proclaimed that the cost of environmental degradation to the Chinese economy had by 2010 rocketed threefold since 2004, to about $230 billion, or 3.5 percent of the nation's gross domestic product.[3] Not all of that is directly related to climate change, for the Chinese have been pretty effective at sending acute pollutants into the country's air, water, and soil. But the droughts in the southwest part of the country, the floods in the northeast, and the growing sophistication of Chinese scientists in studying the phenomena have contributed to China's mounting concern over the impacts of climate change on its economy. They watch warily, too, as the immense river systems emanating from the peaks of the Himalayas, and nourishing much of southern and western China, undergo alterations similar to those experienced in the Sierra Mountains of far-off California. Instead of water falling as snow, the warming atmosphere sends it down as rain, which threatens floods in the winter and not enough water in the dry spring and summer—a potentially disastrous trend, because the Himalayas have served as a kind of frozen reservoir for Southeast Asia, just as the Sierras have for California.

So Brazil and China have their own imperatives to act on climate change, and they are doing so in ways that are sometimes parallel with but do not intersect our own. Each country is both trying to deal with the consequences of climate change and trying to slow it down, fitful as the efforts may be, and they are taking their action largely independent of the international community. "China is devising a carbon price on their own," commented Orville Schell, director of the Center on US–China Relations at the Asia Society and a veteran of reporting on China for *The New Yorker*.[4] "They did not want to be forced by some Kyoto or other accord, they've always resisted having some big legal target foisted on them by the 'first world.' They're doing it themselves."

As for the Brazilians, during the waning days of the Copenhagen negotiations in 2009, President Lula, as he is known, sat with his

Brazilian delegation watching President Obama's remarks on closed-circuit TV. The hope in Copenhagen was that the United States and other developed countries would commit to specific emission limits, and major developing countries like China, Brazil, and India would commit to less specific but internationally verifiable reductions. You could almost feel the air being sucked out of the catacomb of corridors in the negotiating center as Obama retreated from any emission reduction commitment by the United States, and threw some of the responsibility for his position back on China's unwillingness to accept international emission monitors operating within its borders. Lula, according to several members of the delegation who were present with him during the speech, was disappointed and infuriated at Obama's dodge and tossed out the speech in which he'd planned to celebrate the hoped-for US commitment to a global accord. Instead, Lula delivered an off-the-cuff rhetorical blaze in which he excoriated the United States and stressed that Brazil was not asking for money from anyone but instead would make its own commitment to reduce emissions—by 39.1 percent from 1990 levels by 2020. That goal would largely be accomplished through the government's assault on deforestation. The sophisticated satellite monitoring operation I'd seen back in Brasilia was a direct outgrowth of that commitment. It was also one of the early signs of chaotic improvisation, in the absence of global agreement, that has reigned to this day.

At every climate summit, the positions have been repeated: The United States will not commit to a global emissions accord in the absence of a Chinese commitment to reduce their emissions, and the Chinese claim the same—a catch-22 that has served both sides over the past nearly two decades of jousting over climate policy. "For quite some time," commented Ma Jun, director of the Institute of Public & Environmental Affairs, an environmental NGO in Beijing, "we have used each other as an excuse for non-action."

Now that rigid Kabuki dance, in which each side behaves predictably along long-established lines, is being broken—shifts that are sporadic and uncoordinated and nowhere close to what is needed, but challenging the established order in the process.

The dichotomy no longer holds. The Kabuki moves are being short-circuited as the economic costs from climate change rise, coming at us in dissonant bursts.

Climate change is not only disrupting the atmosphere and life here on earth, it is upending the geopolitical architecture that has long reigned between developed and developing countries, relationships often characterized by patterns of extraction, on the one hand, and financial assistance on the other. The rules of influence and power are being rewritten. A country like China or Brazil can no longer purport to being, when it comes to climate responsibility, on the same level as, say, Cameroon or Paraguay, just as the United States can no longer purport that its unwillingness to sign on to a global climate accord is due to China's inaction.

It's also rewriting the rules of how we've come to understand traditional economics. As we head toward an economy that places a price on carbon, answering the questions *who pays, and how?* requires jumping over existing ideas of profit, loss, risk, and how to account for it all.

If London was the center of the post-carbon economy of the early twenty-first century, the transformation for the rest of the century will have centers all over the world—from Brussels to Beijing to San Francisco, New York and Brasilia and New Delhi, as well as London, a confusing but intersecting network that has at its heart the clumsy fashioning of a carbon price, little by little etched into the economy.

You could inject a black dye representing carbon into the circulatory system of the twenty-first-century economy and see it appear behind every major calculation from here on out, rising with intensity and focus. The United Nations Environment Program reports that more than two dozen countries now have at least a minimal price for carbon—ranging from all the Kyoto Protocol signatories to Korea, Mexico, the Canadian provinces of Quebec and Alberta, and even India, which slapped a small tax on coal in 2010. In the

United States, that cost comes primarily from California and the northeastern states, which subject energy utilities and refiners to a limited cap-and-trade system, with a cap that is expected to decline 2.5 percent each year from 2015 to 2020. On a federal level, the United States has, as yet, little influence over how this price is created, but Americans are paying it nevertheless through higher prices, measured at this stage in very tiny increments, paid for in our imports, little by little, from locales that do apply a cost for the emission of greenhouse gases.

From Europe to China to the United States, and all those places in between, however, the cost of carbon has been so minimal as to be largely ineffectual in triggering a change in investment priorities. The Chinese initiative, while of immense symbolic and precedent-setting significance, is really quite small in the face of the country's onslaught of emissions—just as cap and trade in Europe has not come close to succeeding in leveraging a price that would lead to significant shifts in energy priorities. The revolution in clean energy investment that architects of carbon pricing hope to trigger has yet to arrive. A considerable rewriting of the economic rules is, still, in order.

Before Alastair MacGregor became the chief operating officer of the environmental accounting firm Trucost, he managed investments for a London mutual fund, a fund that followed traditional accounting practices and paid little heed to companies' contribution to climate change. "We had a saying back then," he said. "You'd never get fired for following one of the big companies, like Putnam or Fidelity. Because if they go bust, everybody's going bust. You're not singled out." In other words, follow the pack.

Now MacGregor specializes in identifying the often unseen environmental risks behind the fossil fuel industries and the financial prospects for companies shifting away from them. The challenge is to alter the calculus so that "the pack" has at least equal impetus to consider the move toward less destructive investments. He explained that the value of a company is based roughly half on its current income-generating potential. The other half, he said, "is based on its long-term prospects, its value over the next fifty years." Carbon without a significant cost does not change that underlying calculation. Betting against carbon, the "dark spread," and in favor of fossil fuel alternatives, the "clean dark spread," needs to become the safe bet.

Currently, the risks from the dark spread are largely shielded from public view. Their hidden costs prevent alternatives from being judged against traditional market criteria. MacGregor called this "the illusion that their revenues represent an added value to society. You must also subtract their costs to society." Trucost calculated that the climate-based damage to the planet's natural capital from the world's top oil and coal companies, utilities, and steel and cement manufacturers could amount to "over 50% of the companies' combined earnings."[5] Just ninety of those fossil-fuel-intensive companies are responsible for two-thirds of the emissions on the planet.[6]

Such companies in the United States have few requirements to disclose the environmental costs and risks of their production. In 2010, the Securities and Exchange Commission issued a voluntary guidance, suggesting that companies begin to report the risks from climate change in 10-K forms outlining their financial status. The results, with few exceptions, have been largely pro forma, according to Ceres, the network of finance institutions and public interest groups trying to steer investments in a more sustainable direction, which analyzed the disclosure forms.[7] Ceres concluded that there had been "little improvement" from companies' negligible reporting of climate risks dating from before the guidance. So we might consider these pages an alternative 10-K, describing, in narrative form, the convergence of financial and environmental risks triggered by climate change.

By contrast, reporting of greenhouse gas emissions is now mandatory on the French and British stock exchanges, which are seen as
test cases for the entire European Union. This conjures a scenario
in which transnational companies supply two sets of data: one to
the European exchanges, where they provide investors with their
climate-related risks, and another in the United States where they
are under no obligation to do so.

Externalities are now being looked at in a different light by
economists—not as signs of vigorous productive wealth, but as
generators of costs and signs of inefficiency. Examples of those
inefficiencies are surfacing from the financial markets themselves:
The one hundred companies in *Newsweek*'s yearly Green Rankings,
for example, outperformed the S&P 500 by 4 to 5 percent every year
between 2008 and 2012. "The worst polluters," said Cary Krosinsky,
an environmental economist who helped devise the methodologies
used by the index, "come with real vulnerability and volatility."
Krosinsky formerly worked with the American office of Trucost
and is now executive director of the Network for Sustainable
Financial Markets, which is working to compel greater attention to
environmental risks in the financial markets.

"Companies that are most energy efficient are outperforming their
competitors which are not," he said. "Highly-polluting activities can
affect shareholder value because they . . . come back as insurance
premiums, [punitive] taxes, inflated input prices and the physical
cost associated with disasters, costs which can reduce future cash
flows and dividends."

These factors are beginning to creep into the insular trading
posts where investment decisions are made. At the end of 2013,
Bloomberg, which supplies the terminals that provide minute-by-
minute performance data on the world's publicly traded companies,
added an extra feature for its more than thirty thousand subscribing
financial traders: The Carbon Risk Valuation Tool presents a numer-
ically translated portrait of the numerous and unofficial risk factors
of investments in oil companies and other fossil-fuel-intensive

industries. For the first time, potential investors can toggle between the two sides of the bet on fossil fuels: Choose one tab and you get their financial performance; choose another and you get their greenhouse gas emissions and an assessment of the risks that new regulations could leave their assets untapped and underground. "We're seeing a changing of the norms," commented Curtis Ravenel, head of the Global Sustainability Group at Bloomberg in New York. "How companies manage environmental issues is a proxy for how well a company is managed. We believe the day will come when you will have full-cost accounting, with more and more companies taking a longer view."

There is also the big payout liability factor, identified by the global insurance giant Lloyd's of London, of companies being held accountable for the economic consequences of climate change. "We foresee," the company concluded in 2012, "an increasing possibility of attributing weather-related losses to man-made climate change factors. This opens the possibility of courts assigning liability and compensation for claims of damage." Those risks are getting more intense by the day.

The United Nations Environment Program warns that the lack of a uniform climate price and uniform regulatory standards, a feature of the chaotic situation we face today, creates an added risk: When extreme events happen, "the current lack of policy ambition on climate change will likely lead to more sudden and radical policy interventions in the future." In other words, a more substantive commitment now could head off far more economically disruptive responses later. As Mark Trexler of the consulting firm The Climatographer puts it: "What happens to fossil-fuel companies when you have two Katrinas and twenty-three tornadoes and the world wakes up and finally imposes a meaningful price on carbon? The comparative balance sheets will look very different."

So pricing carbon is not some abstract math exercise. The math behind the price is also ultimately grounded in a more monumental calculation, where math meets science meets the global economic order, which faces what has been characterized as the "carbon bubble"—the Damocles Sword that hangs above all our heads. The "bubble" refers to the vast quantity of potential carbon emissions now buried in the ground in oil and gas reserves controlled by fossil fuel companies that are poised to be extracted over the coming decades.

The London-based Carbon Tracker Initiative, working with the London School of Economics, used IPCC data to devise a "carbon budget" of some nine hundred gigatons of carbon dioxide emissions—roughly a trillion tons—an allocation of greenhouse gases over the coming forty years that would give us an 80 percent probability of keeping temperatures from rising more than 2 degrees Celsius above pre-industrial levels.[8] Beyond that, scientists warn, the impacts of climate change will accelerate and intensify, perhaps unstoppably. As of 2012, we'd already used up about five hundred billion tons, which leaves us with another half a trillion to burn by 2050. ExxonMobil and Russia's Lukoil[9] could release forty gigatons of carbon dioxide into the atmosphere—almost one-tenth of the carbon budget from two companies alone—if they exploit every one of their underground and underwater reserves. A precautionary approach to our remaining carbon budget would permit just 20 percent of those reserves to be pumped to the surface.

What happens to all that oil and gas and coal left untouched? They become what the Carbon Tracker dubbed "stranded assets," an unmonetized albatross around the neck of the fossil fuel companies. But that's only if there is enough of a penalty on greenhouse gases to make it economically infeasible to extract those fossilized resources rather than simply leaving them there to decay for another millennia or two, or three. If not, given the imperatives of the energy marketplace, they'll be pumped up, sold, and processed for our energy utilities, our refineries, our gas tanks, and we'll exceed that carbon budget pretty quickly. Annual investments in low-carbon energy need to double by 2020, and double again by 2050, according to the International Energy Agency, in order to avoid reaching that two-degree trigger.

Behind the multiple moving parts of the extraordinarily complex carbon markets—the validators, the offset developers, the carbon traders, the new commodity, the amazing machinations of a multi-billion-dollar market created out of thin, CO_2-saturated air—has been the effort to give carbon the real-cost price that would encourage such investments. There is a reason why the Europeans, for example, have hung in there so long with that orphaned American invention—the hope that ultimately the price will reach a level that will spur an actual transformation in energy-investment patterns. "We want to be in position so that when the world wakes up and realizes how serious a problem this is and wants to get serious about moving toward a post-carbon world, Europe is ready with the technology," commented Niels Vogelsang, legislative aide to one of the European Parliament's top leaders on climate change, Dan Jørgensen. It's fair to say that the world may not yet have woken up but it's definitely no longer asleep.

At the climate summit scheduled in Paris in 2015, when a successor to the Kyoto Protocol will be debated—and for which the 2014 summit in Lima, Peru, is preparation—the world is expected to review the performance of the carbon markets thus far, looking for ways to tighten the rules and devise additional means of funding the energy transformation needed. The actions of China, and Brazil, the efforts of Europe—and even the efforts of the United States insofar as President Obama has tried to bring some coherence to US climate initiatives through executive action—suggest that these gatherings may not offer the same Kabuki performances as they have in the past.

It will certainly be an opportunity to assess almost a decade of experimentation with cap and trade and the carbon markets they spawned. That convoluted apparatus was designed to sidestep what the Americans had feared back in Kyoto—a price for carbon that would actually have bite, that would actively discourage investments in fossil fuels. In practice, to ensure that the markets consistently perform according to their environmental rationale requires intervention to backstop the price above a meaningful level and guarantee that emission reductions paid for equate to emission reductions achieved. In other words, it would require far more resources for enforcement on a global scale and a price floor that would add up to the very thing that cap and trade was designed to avoid—a carbon tax.

Talk of taxes is of course uncomfortable; few of us, save for the occasional billionaire or multimillionaire like Warren Buffett or Barack Obama, see the need to pay more in taxes. But the idea is starting to seem less and less radical as a means of ensuring not only a meaningful price for carbon but one that offers predictability to businesses, enabling them to plan for future energy sources and expenditures. Ireland, Sweden, Australia, and the Canadian province of British Columbia have already introduced a carbon tax. In 2013, Britain introduced a price floor for coal-fed power stations intended to ensure that their price of emissions, plunging on the carbon markets, reaches at least thirty dollars a ton. In Ireland, a tax imposed in 2010 based on fossil fuel use in homes, offices, and vehicles sparked a surge in sales of more fuel-efficient cars and a 6.7 percent drop in the country's emissions, even as the economy grew slightly, which helped finance the creation of thousands of new green jobs. In Australia, a twenty-three-dollar-per-ton tax on the country's top five hundred producers of greenhouse gases has generated more than thirty-nine billion dollars in new revenues, five billion dollars of which went to supporting new technologies to retool the economy toward renewable energy sources, and more than a third of which was returned to low- and moderate-income consumers to offset higher fuel costs. The government, which estimated that the country's emissions dropped by 7 percent due partly to that initiative, claimed that it would position the country for the low-carbon economy taking shape. "Those in the vanguard of clean energy now will make a killing in the future," said Karen Lanyon, the Australian consul general in Los Angeles.

A concept gaining traction in Europe—it's been endorsed by twelve members of the EU—is for a tax on financial transactions. A levy of just 0.05 percent on each purchase of stocks, bonds, and derivatives could raise up to $450 billion that could be channeled into the UN's Green Climate Fund—an initiative that is supposed to help finance climate mitigation and adaptation in developing countries but as yet has few sources of actual funding.

However—and that's a hedge that virtually every major economist uses when speaking about the relative efficiency of a tax versus cap and trade—the very word *tax* is enough to send most members of the US Congress, not to mention most Americans and taxpayers around the world, running for cover—as it did Al Gore in 1997. Six months after I spoke with Lanyon, her boss was out of a job: Australia's Labour government lost the national election, and the new prime minister Tony Abbott set about dismantling a significant portion of the Australian government's climate policies. Needless to say, China's ability to act decisively is largely because it doesn't face the same popular pressures as someone like Walter De Simoni does in democratic Brazil.

Nevertheless, even in the United States, the idea that was considered and rejected during the Kyoto negotiations is quietly taking hold across the ideological spectrum as the most efficient way to avoid the flaws of the market and provide a direct cost to the use of fossil fuels. On the conservative side, former US representative Bob Inglis of South Carolina, now head of the Energy and Enterprise Initiative at George Mason University, argues for a tax on coal that echoes with the language of Trucost and other environmental economists. "Let's make coal fully accountable for all its health costs," he told an interviewer on the environmental web-video show E&E. "If you do that, it really changes the economics . . . We're paying all the costs of coal-fired electricity right now in our insurance premiums and through health care. We're paying all right, just not at the [electricity] meter." Inglis's explanation suggests a common ground where liberals and conservatives might meet—a carbon tax as a way to save money in the long run.

On the liberal side, two US senators, Barbara Boxer of California and Bernard Sanders of Vermont, proposed a bill in 2013 that would impose a twenty-dollar-per-ton price on emissions generated by utilities and oil refineries. The price would rise 5 percent every year over the next decade, bringing it by 2025 to thirty-five dollars. Though it stood little chance of passing the Republican House, the introduction of the bill suggested that the debate was beginning to move toward a more direct penalty on carbon than that offered in the markets. Critically, 60 percent of the revenues generated would be devoted to rebates for consumers. That's because there's no way around the fact that a price for carbon means a higher price for energy and gasoline.

Indeed, there are fundamental matters of equity raised by the effort to generate a price for carbon. The central question is, who pays for the extra costs?

"If fossil fuel companies simply raise their prices, thus increasing their profit margins and shunting the price onto consumers, that's a fundamentally unequal distribution of the burden," said Donald Browne, a professor at the Widener Law school, who has been at the forefront of efforts to integrate ethical considerations into the climate response. "We need to have a morally relevant criteria for distributing the carbon budget."

If the costs of carbon are simply passed directly on to consumers, then the oil companies and other intensive fossil-fuel users will continue to profit from the same misleading cost accounting they've benefited from for more than a century now, and leave the public to feed their existing margins. "We need a way to ensure that the burden of increased energy costs does not fall disproportionately on people with less resources to pay," commented Peter Barnes, founder of the Working Assets phone company and one of the first to propose the fee-and-dividend concept to ensure that taxpayers are not left alone with the burden of covering increased costs for energy.[10] Fees would be paid on greenhouse gases at points of extraction or points of entry into the country based on their greenhouse gas load, and rebates provided to consumers in the form of tax breaks or rebate checks.

Many businesses, in Europe and America, are coming round to the idea that a tax is the most direct way to establish a predictable price for carbon that would enable them to plan long-term and create a new level playing field, one that starts at least with a more accurate rendering for the costs of energy. It would, finally, give businesses, investors, the government, and citizens a clear eye on the ledger of profit and loss.

Ultimately that ledger gets down to the common resource of our atmosphere and the consequences of its disruption down here on earth. A basic underpinning to that idea came first from another British economist, William Forster Lloyd, who observed in the mid-nineteenth century the corrosive results when individuals pursued their own economic interests without considering the costs to commonly held resources. That observation, delivered in a series of lectures at Oxford University, became what we now refer to as "the tragedy of the commons"—for which climate change is the penultimate example.

A panoramic shift in thinking about energy is under way, creating new imperatives of financial decision making, reshaping geopolitical relations, giving rise to new powers and diminishing others, including the United States, as the world struggles to respond to the looming specter of catastrophic climate change. As the financial analysts at PriceWaterhouseCoopers put it: "A low carbon economy race is on, and businesses, of any size, cannot ignore it. Our generation's industrial revolution will change business practices, rethink energy and resource consumption and . . . drive new technologies to emerge quicker than ever before."[11] At an LSE-sponsored gathering in San Francisco, Nicholas Stern put it similarly: "We are in a green race to the top," he said. "Those who don't change their practices to adopt to the social costs of production will in the long run lose." Climate change heightens just how much our economic future is intertwined with our environmental future.

Steadily, slowly, the world is inexorably moving toward a price for the primary pollutant of our time. How much and when are of course key questions. But the very process under way has already been as disruptive as any new technology has ever been. As the world wakes up, who will be where as the deck shuffles?

A significant and globally relevant price for carbon would have immediate impact on, for example, the prospects of green industries clustering in cities like Manchester, England, and Pittsburgh, Pennsylvania, and those starting to emerge in Shenzhen in China's Guangdong province. For the developers of solar and wind power around the world, it will finally offer them a level playing field with which to compete with the makers of fossil fuels. It would encourage farmers to reduce their reliance on fossil-fuel-intensive inputs, and tend to their soils in ways that build in long-term resilience to the changes to come. It would reduce the number of oil spills we let loose each day into the atmosphere. It would end the distortion that has been a dominant feature of the so-called free market, in which profits generated in the fossil-fuel-heavy private sector are built on risks assumed by the public. It would contribute to expanding the parallel green economy that is already generating new forms of employment and innovation.

It would, in short, transform Branca Americano's low-carbon coffee cup from the eco-exception in the global economy to the norm.

Acknowledgments

First, I would like to express my deep gratitude to the funders who provided the financial support that enabled me to pursue the reporting and writing of this book. Thank you to Diana Cohn, executive director of the Panta Rhea Foundation, whose steadfast support, friendship, and faith have been central to its completion; and to Leslie Leslie and Annette Gellert, whose Fred Gellert Family Foundation has provided a vital backbeat to my work over the years. Also, thanks to Sarah Bell and the 11th Hour Project, whose interest in how farmers are contending with climate change helped to sustain my reporting forays into the Central Valley. It couldn't have happened without them.

I was based at the Center for Investigative Reporting during the genesis of this project; thanks to Executive Director Robert Rosenthal for that early support. I am grateful to Cherilyn Parsons and Sharon Tiller, who provided useful comments and critiques while at CIR. Former CIR reporter Sarah Terry-Cobo provided background reporting to the chapters on forests and oil. Interns Alexandra Perloff-Giles and Christopher Holm also contributed research. Many thanks to the Mesa Refuge, which provided a beautiful and inspiring perch to make the switch from reporting to writing.

Deborah Kirk provided valuable editorial advice, as did my multi-talented agent Diana Finch. Peter Cunningham, Jackie Bennion, Mark Hertsgaard, and Phillip Frazer were critical readers and listeners along the way. At Chelsea Green, Joni Praded was a pleasure to work with as she turned her sharp eye onto this tale; and thanks to publisher Margo Baldwin, for seeing early on what I tried to do and sticking by it.

Thank you to my family—Erik, Seth, Avi, Shayna, Abbey, Lani—for being there. And to Zoe Fitzgerald Carter, whose pleasure with words and melody enriched the creation of this book.

Notes

Introduction

1. Milly, P. C. D., Julio Betancourt, et al., "Stationarity Is Dead: Whither Water Management?" *Science* 319, February 1, 2008.

2. "Warming of the Oceans and Implications for the (Re)insurance Industry," Geneva Association of Risk and Insurance Economics, June 2013.

3. "Universal Ownership: Why Environmental Externalities Matter to Institutional Investors," United Nations Environment Program Finance Initiative, 2011.

4. "World Energy Outlook 2013 Factsheet," International Energy Agency, www.iea.org/media/files/WEO2013_factsheets.pdf.

5. Testimony of Howard Shelanski, administrator for the Office of Information and Regulatory Affairs, US Office of Management and Budget, to the House Committee on Oversight and Government Reform, Subcommittee on Energy Policy, Healthcare and Entitlements, July 18, 2013.

6. "Inventory of US Greenhouse Gas Emissions and Sinks: 1990–2012," Environmental Protection Agency.

7. "World Energy Outlook: Energy Subsidies," International Energy Agency, 2014. www.iea.org/publications /worldenergyoutlook/resources/energysubsidies.

8. Heede, Richard, "Tracing Anthropogenic Carbon Dioxide and Methane Emissions to Fossil Fuel and Cement Producers, 1854–2010," *Climatic Change* 122, January 2014.

1: Dogfight Over My Flight

1. The ICAO Carbon Emissions Calculator can be found here: www.icao.int/environmental-protection/CarbonOffset/Pages/default.aspx.
2. "Aviation and the European Union's Trading Scheme," Congressional Research Service, June 11, 2012.
3. Testimony by Annie Petsonk, International Counsel for the Environmental Defense Fund, to the US Senate Committee on Commerce, Science and Transportation, June 6, 2012.
4. Burkhardt, Ulrike, Bernd Karcher, and Ulrich Schumann, "Global Modeling of the Contrail and Contrail Cirrus Climate Impact," American Meteorological Society, April 2010.
5. This calculation comes from the independent trade journal *GreenAir Online*, which reports on the aviation industry: www.greenaironline.com/news.php?viewStory=217.
6. Keen, Michael, and Jon Strand, "Indirect Taxes on International Aviation," *Fiscal Studies* 28, no. 1, 2007.
7. Grabar, V. A., M. L.Gitarskii, et al., "Assessment of Greenhouse Gases Emission from Civil Aviation in Russia," *Russian Meteorology and Hydrology* 36, no. 1, 2011.
8. I wrote a story on this legal case as it was unfolding for *The Atlantic*: "Green War in the Skies: Can Europe Make US Planes Pay for Pollution?," October 5, 2011. www.theatlantic.com/business/archive/2011/10/green-war-in-the-skies-can-europe-make-us-planes-pay-for-pollution/246105.
9. "Oral Submissions on Behalf of the Air Transport Association of America (ATA), United Continental Airlines, & American Airlines," Case C-366/10, Grand Chamber Hearing, European Court of Justice, July 5, 2011.
10. "ICAO and Aviation Emissions: The Clock Is Ticking" (timeline of dispute), Transport & Environment, February 4, 2013.

11. "Written Observations of the Second Interveners," Case C-366/10, European Court of Justice, published by the court on October 20, 2010.
12. "Judgment of the Court (Grand Chamber)," European Court of Justice, case C-366/10, December 21, 2011.
13. "Joint Declaration of the Moscow Meeting on Inclusion of International Civil Aviation in the EU-ETS," February 22, 2012, www.ruaviation.com/docs/1/2012/2/22/50; also see Kramer, Andrew E., "Opponents to European Airline Emissions Law Prepare Their Countermeasures," *New York Times*, February 22, 2012.
14. Malina, Robert, Dominic McConnachie, et al., "The Impact of the European Union Emissions Trading Scheme on US Aviation," *Journal of Air Transport Management* 19, 2012.
15. Schwartz, Moshe, Katherine Blakeley, et al., "Department of Defense Energy Initiatives: Background and Issues for Congress," Congressional Research Service, December 10, 2012.
16. "Beginner's Guide to Aviation Biofuels," Air Transport Action Group, May 2009.
17. A substantive summary of the EU's retreat is provided by the Stockholm Environment Institute, "The Conflict Over Aviation Emissions," February 2014.

2: Eat, Drink, Pray

1. Boering, K. A. et al., "Trends and Seasonal Cycles in the Isotopic Composition of Nitrous Oxide since 1940," *Nature Geoscience*, 5, April 2012.
2. Luedeling, Elke, Minghua Zhang, et al., "Climatic Changes Lead to Declining Winter Chill for Fruit and Nut Trees in California During 1950–2099," *PLOS ONE* 4, no. 7, July 2009.
3. "Indicators of Climate Change in California," Office of Environment Health Hazard Assessement and Environmental

Protection Agency, State of California, August 2013; interview with Dennis Boldocchi, professor of bio-meteorology and expert on fog, at College of Natural Resources, University of California–Berkeley, August 30, 2012.

4. Some of the Central Valley reporting for this chapter was conducted initially for a documentary, *Heat and Harvest*, that aired on San Francisco's PBS station KQED. (Thanks to Serene Fang, who produced it as we traveled through the valley.) A two-part newspaper series I wrote for the Center for Investigative Reporting accompanied the documentary. All can be seen here: www.centerforinvestigativereporting.org /projects/heat-and-harvest.

5. Lobell, David B., and Christopher B. Field, "California Perennial Crops in a Changing Climate," *Climatic Change*, November 2011.

6. "Climate Change Consortium for Specialty Crops: Impacts and Strategies for Resilience," California Department of Food and Agriculture, 2013.

7. "Regional Water Management Planning with Climate Change Adaptation and Mitigation," Environmental Protection Agency, Region 9, Water Division, 2011.

8. Costello, Christopher J., Olivier Deschenes, and Charles Kolstad, "Economic Impacts of Climate Change on California Agriculture," California Climate Change Center, California Energy Commission, August 2009.

9. "Climate Change and Agriculture in the United States: Effects and Adaptation," US Department of Agriculture, February 2013.

10. "Final Delivery Reliability Report," California Department of Water Resources, June 2012.

11. "Regional Water Management Planning with Climate Change Adaptation and Mitigation," Environmental Protection Agency, Region 9, Water Division, May 11, 2011.

12. "Comparing Futures for the Sacramento–San Joaquin Delta," Public Policy Institute of California, 2008.

13. Public Policy Institute of California and Center for Watershed Sciences, "Delta Hydrodynamics and Water Salinity with Future Conditions: Technical Appendix C," University of California–Davis, July 2008.

14. Howitt, Richard, Jonathan Kaplan, et al., "The Economic Impacts of Central Valley Salinity: Final Report to the State Water Resources Control Board," University of California–Davis, March 20, 2009.

15. Alcamo, Joseph, and Jorgen E. Olesen, *Life in Europe Under Climate Change*, Wiley-Blackwell, 2012.

16. Gillis, Justin, "Climate Change Seen Posing Risks to Food Supplies." *New York Times*, November 1, 2013.

3: Earth, Wind, and Heat

1. "High Risk Series: An Update," Report to Congressional Committees, Government Accountability Office, February 2013.

2. "Agricultural Exposure to Water Stress," World Resources Institute, www.wri.org/applications/maps/agriculturemap.

3. "2013 RMA Crops' Indemnities (as of 02/10/2014)," Risk Management Agency, US Department of Agriculture, www.rma.usda.gov/data/indemnity/2014/021014map.pdf.

4. Shields, Daniel A., "Federal Crop Insurance: Background and Issues," Congressional Research Service, December 13, 2010.

5. "Standard Reinsurance Agreement Final Draft Fact Sheet," Risk Management Agency, US Department of Agriculture, June 10, 2010.

6. Beach, Robert H., Chen Zhen, et al., "Climate Change Impacts on Crop Insurance," Final Report for the Risk Management Agency, US Department of Agriculture, May 2010.

7. "Climate Change and Agriculture in the United States: Effects and Adaptation," US Department of Agriculture, February 2013.

8. "360 Risk Project: Catastrophe Trends," Lloyd's of London.

9. "Soil Matters: How the Federal Crop Insurance Program Should be Reformed to Encourage Low-Risk Farming Methods with High-Reward Environmental Outcomes," Natural Resources Defense Council Issue Paper, August 2013.

10. Lobell, David B., Angela Torney, and Christopher B. Field, "Climate Extremes in California Agriculture," *Climatic Change*, November 24, 2011.

11. I addressed some of the issues surrounding genetically modified seeds in *Sowing Disaster?* for *The Nation* (October 28, 2002) and the PBS documentary series *NOW with Bill Moyers* and later in the book *Exposed*.

12. "The Farming Systems Trial, Celebrating Thirty Years," Rodale Institute.

13. Davis, Adam S., Jason D. Hill, et al., "Increasing Cropping System Diversity Balances Productivity, Profitability and Environmental Health," *PLOS ONE*, October 10, 2012; United Nations Conference on Trade and Development, "Wake Up Before It's Too Late: Make Agriculture Truly Sustainable Now for Food Security in a Changing Climate" (compilation of reports and studies), *Trade and Environment Review 2013*, September 2013.

14. In my previous book, *Exposed: The Toxic Chemistry of Everyday Products and What's at Stake for American Power* (Chelsea Green, 2007), I explored the implications of Europe's and America's differing environmental health standards.

15. "Product Carbon Footprint Summary," TESCO, August 2012.

16. www.ewg.org/meateatersguide/eat-smart.

17. Jones, Christopher, Daniel Kammen, and Daniel McGrath, "Consumer-Oriented Life Cycle Assessment of Food, Goods and Services," Berkeley Institute of the Environment, UC Berkeley, March 3, 2008.

18. Ripple, William J., Pete Smith, et al., "Ruminants, Climate Change and Climate Policy," *Nature Climate Change* 4, January 2014.

4: The Forest for Its Carbon

1. Tufts University, Office of Sustainability. http://sustainability .tufts.edu/carbon-sequestration. The Tufts study summarizes Department of Energy guidelines contained in the DOE's "Sector-Specific Issues and Reporting Methodologies Supporting the General Guidelines for the Voluntary Reporting of Greenhouse Gases under Sections 1605(b) of the Energy Policy Act of 1992."
2. "Methods for Calculating Forest Ecosystem and Harvested Carbon with Standard Estimates for Forest Types of the United States," US Department of Agriculture, Technical Report NE-343, April 2006.
3. Schapiro, Mark, and Sarah Terry-Cobo, "Timber Companies Stand to Benefit from New Climate Law," California Watch/ Center for Investigative Reporting, December 15, 2010, http://californiawatch.org/dailyreport/timber-companies -stand-benefit-new-climate-law-7469.
4. I journeyed to Manaus for a documentary produced for the PBS newsmagazine *FRONTLINE/World*. I want to acknowledge the invaluable contribution of colleagues Andres Cediel, producer, and Daniela Broitman, associate producer, who joined me on this journey and produced the segment "The Carbon Hunters," which aired nationwide on PBS.
5. "Carbon Sequestration in Forests," Congressional Research Service, August 6, 2009.
6. Some of the material from this section on Guaraqueçaba is drawn from my article "GM's Money Trees," *Mother Jones*, November–December 2009.
7. "Covering New Ground: State of the Forest Carbon Markets 2013," Ecosystem Marketplace, 2013, www.forest-trends.org /documents/files/SOFCM-full-report.pdf.

8. "The Impact of EU Consumption on Deforestation: Comprehensive Analysis of the Impact of EU Consumption on Deforestation," study funded by the European Commission, DG-Environment, Technical Report 2013-063; "Trade Emerging as a Key Driver of Brazilian Deforestation," *Science Daily*, April 4, 2013.

9. "Mobilising International Climate Finance: Lessons from the Fast-Start Finance Period," Overseas Development Institute and World Resources Institute, November 2013.

5: Carbon in the Tank

1. I reported a short documentary on the *Prestige* oil spill, "The Lawless Sea," for the PBS newsmagazine *FRONTLINE/World*, which provided the basis for the reporting on the *Prestige* in this chapter. It can be viewed at: www.pbs.org/frontlineworld /stories/spain.

2. *The Price of Gas*, reported by Sarah Terry-Cobo and produced by Carrie Ching, Center for Investigative Reporting, http://centerforinvestigativereporting.org/reports/price -gas-2447. Thanks to Sarah Terry-Cobo for some of the research on the environmental footprint of gasoline in this chapter.

3. "Taxation, Innovation and the Environment: A Policy Brief," Organisation for Economic Co-operation and Development, September 2011.

4. Michalek, Jeremy, Mikail Chester, et al., "Valuation of Plug-In Vehicle Life-Cycle Air Emissions and Oil Displacement Benefits," *Proceedings of the National Academy of Sciences (PNAS)* 108, no. 40, October 4, 2011.

5. "Profits and Pink Slips: How Big Oil and Gas Companies Are Not Creating US Jobs or Paying Their Fair Share," House of Representatives Committee on Natural Resources, Democrats (Minority Report), September 8, 2011.

6. "Inventory of Estimated Budgetary Support and Tax Expenditures for Fossil Fuels 2013," Organisation for Economic Co-operation and Development, January 28, 2013, www.oecd .org/site/tadffss.

7. Pfund, Nancy, and Ben Healey, "What Would Jefferson Do? The Historical Role of Federal Subsidies in Shaping America's Future," DBL Investors, September 2011.

8. "Oil & Gas," Open Secrets: Center for Responsive Politics, www.opensecrets.org/industries/totals.php?ind=E01.

9. Whitley, Shelagh, "Time to Change the Game: Fossil Fuel Subsidies and Climate," Overseas Development Institute, November 2013.

10. "Energy Subsidy Reform: Lessons and Implications," International Monetary Fund, January 28, 2013.

11. "Greenhouse Gas Reporting Program 2012 Reported Data," Environmental Protection Agency. www.epa.gov/ghgreporting /ghgdata/reported/index.html.

12. http://cleantechnica.com/2013/09/29/massive-growth -electric-cars-us-drives-electric-cars-infographic.

13. "International Comparison of Light-Duty Vehicle Fuel Economy: An Update Using 2010 and 2011 New Registration Data," Working Paper 8, Global Fuel Economy Initiative, International Energy Agency.

14. "Environmental Assessment of Plug-In Hybrid Electric Vehicles," Final Report, Electric Policy Research Institute and Natural Resources Defense Council, July 2007; "State of Charge: Electric Vehicles' Global Warming Emissions and Fuel-Cost Savings Across the United States," Union of Concerned Scientists, June 2012.

15. Hilzenrath, David S., "Oil Rigs' Safety Net Questioned as Governments Rely on Private Inspections," *Washington Post*, August 15, 2010.

16. "*Deepwater Horizon*: Coast Guard and Interior Could Improve Their Offshore Energy Inspection Program," testimony

before the House of Representatives Subcommittee on Coast Guard and Maritime Transportation, statements by Stephen Caldwell and Frank Rusco, Government Accountability Office, November 2, 2011.

6: A Tale of Three Cities

1. Hoesly, Rachel, Mike Blackhurst, et al., "Historical Carbon Footprinting and Implications for Sustainability Planning: A Case Study of the Pittsburgh Region," *Environmental Science & Technology*, March 29, 2012.
2. "Pittsburgh Climate Action Plan 2.0," Pittsburgh Climate Initiative, 2012.
3. "Pittsburgh Greenhouse Gas Inventory," Pittsburgh Climate Initiative, October 2010.
4. Peters, Glen, Jan Minx, Christopher Weber, and Ottmar Edenhofer, "Growth in Emissions Transfers via International Trade from 1990 to 2008," *Proceedings of the National Academy of Sciences (PNAS)* 108, no. 21, May 24, 2011.
5. Wang, H., R. Zhang, et al., "The Carbon Emissions of Chinese Cities," *Atmospheric Chemistry and Physics* 12, 2012.
6. Ross, Andrew, *Fast Boat to China: Corporate Flight and the Consequences of Free Trade*, Pantheon Books, 2006.
7. Davis, Steven U., and Ken Caldeira, "Consumption-Based Accounting of CO_2 Emissions," *Proceedings of the National Academy of Sciences* 107, no.12, March, 2010.
8. The Museum of Science and Industry (MOSI) in Manchester, UK, offers an insightful display of exhibitions on Manchester's historic role in the industrial revolution, www.mosi.org.uk.
9. Tocqueville, Alexis de, *Journeys to England and Ireland*. 1835.
10. "Manchester: A Certain Future: Our Collective Action on Climate Change," Manchester City Council, December 2009.
11. "UK's Carbon Footprint 1993–2010," UK Department for Environment, Food & Rural Affairs, December 13, 2012.

12. "Consumption-Based Emissions Reporting: Government Response to the Committee's Report of Session 2010–2012— Energy and Climate Change, Appendix: Government Response," House of Commons Committee on Climate Change publication, January 4, 2013.
13. US Court of Appeals for the Ninth Circuit, Opinion re: *Rocky Mountain Farmers Union v Corey*, case no. 12-15131, September 18, 2013.

7: The Clean Dark Spread

1. "Natural Capital at Risk: The Top 100 Externalities of Business," Trucost, April 2013.
2. Determining precise impacts on an economy as enormous and complex as the United States' is a necessarily complex enterprise. The estimate of 2 to 3 percent appears to be an evolving consensus figure; with a topography as varied as that of the US, the impacts vary from region to region. Key source materials on these estimates: Center for Integrative Environmental Research, "The US Economic Impacts of Climate Change and the Costs of Inaction: A Review and Assessment," University of Maryland, October 2007; "Climate Vulnerability Monitor," DARA International, September, 2012; Ackerman, Frank, and Elizabeth A. Stanton, "The Cost of Climate Change: What We'll Pay if Global Warming Continues Unchecked," Natural Resources Defense Council, May 2008; "Draft: National Climate Assessment," US Global Change Research Program, interagency research center, 2013; Davenport, Coral. "The Scary Truth About How Much Climate Change Is Costing You," *National Journal*, February 7, 2013.
3. Estrada, Francisco, Elissaios Papyrakis, et al., "The Economics of Climate Change in Mexico: Implications for National/Regional Policy," *Climate Policy*, August 2013.

4. The World Bank's Carbon Finance Unit issues yearly reports on the status of the global carbon markets: "State and Trends of the Carbon Market Report 2012," "2011," etc.; and, after 2012, "Mapping Carbon Pricing Initiatives."

5. "Greenhouse Gas Equivalencies Calculator," Environmental Protection Agency, www.epa.gov/cleanenergy/energy -resources/calculator.html#results.

6. Hammoudeh, Shawkat, Duc Nguyen, and Ricardo Sousa, "Energy Prices and CO_2 Emission Allowance Prices: A Quantile Regression Approach," Working Paper 06. Center for Political and Economic Research (NIPE), University of Minho, Portugal, 2014.

7. Mengewein, Julia, and Tino Anderson, "Europe Needs Carbon Floor Price to Boost Gas-Fired Plant Profits," *Bloomberg Finance News*, September 5, 2013.

8. Johnson, Laurie T., Starla Yeh, and Chris Hope, "The Social Cost of Carbon: Implications for Modernizing our Electricity System," *Journal of Environmental Studies and Sciences 3*, no. 4, December 2013.

9. Paul Watkiss Associates, "The Social Cost of Carbon," report for the UK Department of Environment, Food and Rural Affairs, 2008.

10. "A Dangerous Obsession" (research report), Friends of the Earth UK, November 2009.

11. "The Colour of Growth: Maximising the Potential of Green Business," Confederation of British Industries, September 2013.

12. "Carbon Capitulation," Deutsche Bank Markets Research, April 16, 2013.

8: Trading in Hot Air

1. The award-winning film *Carbon Crooks*, by the Danish documentary filmmaker Tom Heinemann, chronicles some of these

and other schemes that have afflicted the carbon markets from their inception. More information here: www.carboncrooks.tv.

2. I was invited to INTERPOL after reporting extensively on abuses in the carbon markets, and blogged about the summit at the time for *Mother Jones*, www.motherjones.com /environment/2010/10/interpol-carbon-trading-fraud.

3. "Guide to Carbon Trading Crime," INTERPOL, Environmental Crime Programme, June 2013.

4. "Recommendations for the CDM Reform: End Climate Finance for Coal Power," position paper for COP-19, Warsaw, Center for International Environmental Law and Carbon Market Watch, November 2013.

5. I wrote a detailed investigation of the carbon auditors in "Conning the Climate," *Harper's*, February 2010.

6. "Assessing the Impact of the CDM Development Mechanism," CDM Policy Review Panel, CDM Policy Dialogue, July 2012.

7. Schapiro, Mark, "'Perverse' Carbon Payments Send Flood of Money to China," *Yale Environment 360*, December 13, 2010, http://e360.yale.edu/feature/perverse_co2_payments_send _flood_of_money_to_china/2350.

9: The Coffee and the Cup

1. Margulis, Sergio, and Carolina Burle Schmidt Dubeux, "The Economics of Climate Change in Brazil: Costs and Opportunities," University of São Paulo, Faculty of Economics (FEA), 2011.

2. "The Response of China, India and Brazil to Climate Change: A Perspective for South Africa," Centre for Development and Enterprise, Smith School of Enterprise and the Environment, University of Oxford, England, November 2012.

3. Wong, Edward, "Cost of Environmental Damage in China Growing Rapidly Amid Industrialization," *New York Times*, March 29, 2013.

4. Orville Schell is also coauthor, with John Delury, of *Wealth and Power: China's Long March to the Twenty-First Century*, a substantive assessment of China's political and economic rise.

5. "Universal Ownership: Why Environmental Externalities Matter to Institutional Investors," Trucost report conducted for the United Nations Environment Program Finance Initiative.

6. Heede, Richard, "Tracing Anthropogenic Carbon Dioxide and Methane Emissions to Fossil Fuel and Cement Producers, 1854–2010," *Climatic Change* 122, January 2014.

7. "Reducing Systemic Risks: The Securities & Exchange Commission and Climate Change," Ceres, February 2014.

8. "Unburnable Carbon 2013: Wasted Capital and Stranded Assets," Carbon Tracker Initiative, in collaboration with the Grantham Research Institute on Climate Change and the Environment.

9. McKibben, Bill. "Global Warming's Terrifying New Math." *Rolling Stone*, July 19, 2012.

10. Barnes's book *With Liberty and Dividends for All: How to Save Our Middle Class When Jobs Don't Pay Enough* explores some of the ethical quandaries presented by using the markets to establish a climate price, as well as other matters of economic equity.

11. "How to Assess Your Green Fraud Risks," PriceWaterhouseCoopers, 2011.

Index

DePaul University, 13
Derwent, Henry, 134, 139, 141–42
Deutsche Bank, 128, 142, 144
differentiated responsibility principle, 12–13,
   137, 141
Directorate-General for Climate Action, 21,
   130, 132
Directorate-General for Mobility and
   Transport, 5
discount rate principle, 135–36
distributed energy term, 100–101
DNV, 159, 161, 164
Dorsey, Michael, 146
Dowie, Mark, 73
droughts
   California, 30, 32–33, 37, 38–39, 42
   countrywide, 42
   insurance payouts for, 49–50
Duarte, John Jr., 29–31
Dust Bowl, 46

eCO2market, 155
economic costs
   of climate change, ix–x, xii–xv, 45–46,
      126–28, 134–37
   environmental disasters, 87–91, 103–7
   of financial crime, 151–56
   for greenhouse gas emissions
      (*see* carbon pricing)
   See also New Carbon Economy
Economic Development Administration, 39
economic systems
   carbon markets as
      (*see* carbon markets)
   production- *vs.* consumption-based,
      114–15, 118–22
economy. *See* New Carbon Economy
EcoSecurities, 142, 160
Ekstra Bladet, 154
electric cars, 99–101
Elkjaer, Bo, 152
embedded emissions, 113–15, 120
embodied deforestation concept, 84
embodied emissions, 118
Emirates, 16
emission allowances
   aviation, 4–5, 15, 20–21
   carbon markets and, 133–34, 144–46
   China requiring, 171–72
   Czech registry theft of, 157
   explained, xvii–xviii, 130–31
   regional market experiments with, 147
emission caps. *See* cap-and-trade system

emission limits
   cap-and-trade system of, xvii–xviii
   European Union regulating, 8–12, 19–21
emissions
   California court ruling on, 123–24
   consumption- *vs.* production
      accountability, 114–15, 118–22
   future mitigation costs, 136–37
   jurisdictions and, 122–23
   reducing, 176–78
   regulating in California, 132
   See also aviation emissions; emission
      allowances
end of stationarity term, ix, xv
Energy and Climate Change Committee
   (UK), 121, 126
Energy and Enterprise Initiative, 185
energy commodities. *See* utilities
energy standards, 100–101
energy technologies. *See* renewable energy/
   technology
Engels, Friedrich, 116
Environment, Food and Rural Affairs
   Committee (UK), 121
environmental costs/risks
   disclosing, 178–80
   financial accounting of, 126
Environmental Crime Programme, 153, 155
environmental crimes
   and law enforcement, 53–54, 76–77,
      153–54, 170
   murder analogy of, xviii
   oil industry's contribution to, 106–7
      (*see also* oil spills)
Environmental Defense Fund, 8, 10, 11, 70,
   140, 145
environmental disasters, 87–91, 103–7
Environmental Investigation Agency, 163
environmental organizations. *See specific
   environmental organizations*
Environmental Working Group, 59–60
EPA (CA), 33, 41
EPA (Environmental Protection Agency)
   data, 30, 67, 91. 131–32, 94, 123, 136, 144
   emission reduction guidelines from, 148
erosion, 49
Errotabere, John, 35–37
The EU Emissions Trading System (EU
   ETS), 130
European Climate Exchange (ECX), 128,
   142, 157
European Commission, 143–44, 146
European Court of Justice

About the Author

PETER CUNNINGHAM

Journalist Mark Schapiro explores the intersection of the environment, economics, and political power, most recently as senior correspondent at the Center for Investigative Reporting. His work has been published in *Harper's, The Atlantic, Yale Environment 360, The Nation, Mother Jones*, and other publications. He has reported stories for the PBS newsmagazine *Frontline/World, NOW with Bill Moyers*, and public radio's *Marketplace*, and is the author of *Exposed: The Toxic Chemistry of Everyday Products and What's at Stake for American Power*. His awards include the Society of Professional Journalists Sigma Delta Chi Award, a DuPont Award, the Society of Environmental Journalists Reporting Award, a National Magazine Award, and a Kurt Schork Award for International Reporting. He teaches at the UC Berkeley Graduate School of Journalism. Schapiro lives in the San Francisco Bay Area.

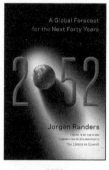